Teaching GCSE Mathematics

This book is dedicated to Debra and Rina with my love.

The quotations which head the chapters are taken from *The Complete Works of Lewis Carroll,* The Modern Library, New York.

Teaching GCSE Mathematics

Zelda Isaacson

HODDER AND STOUGHTON
LONDON SYDNEY AUCKLAND TORONTO

ISBN 0 340 40766 2

First published 1987

Copyright © 1987 Zelda Isaacson

All rights reserved. No part of this publication may be reproduced or transmitted in any form or by any means, electronically or mechanically, including photocopying, recording or any information storage or retrieval system, without either the prior permission in writing from the publisher or a licence, permitting restricted copying, issued by the Copyright Licensing Agency, 7 Ridgmount Street, London WC1E 7AA.

Typeset by Tradespools Ltd, Frome, Somerset
Printed and bound in Great Britain for
Hodder and Stoughton Educational,
a division of Hodder and Stoughton Ltd,
Mill Road, Dunton Green, Sevenoaks, Kent,
by The Eastern Press Ltd, Reading and London

Contents

Acknowledgments	vi
Introduction	5
1 The National Criteria	7
2 The examining bodies and the new syllabuses	21
3 In the classroom	39
4 Classroom examples	74
5 Assessment in GCSE Mathematics	96
Appendix A: A Parents' Workshop	112
Appendix B: Extracts from *The National Criteria – Mathematics*	116
Selected Bibliography	120
Resources List	123

Acknowledgments

Thanks are due especially to the teachers and pupils at the schools named below who contributed classroom examples for this book:

    Cardinal Vaughn R.C. School, Kensington, W14.
    Elizabeth Garrett Anderson School, Islington, N1.
    Haggerston School, Hackney, E2.
    Sir William Collins Secondary School, Camden, NW1.

I am grateful to colleagues at the SMILE centre, the Shell Centre for Mathematical Education and the GAIM project for providing information and material for reproduction.

Rosilinde Scott-Hodgetts insisted that I write this book. Her continuing support and encouragement were invaluable and without her help it would undoubtedly have been a poorer book.

I also wish to thank Alan McLean, Deanne Reynoldson and Anne Walton for reading the manuscript and making helpful comments, and my daughter Rina for helping in other ways.

Any errors or omissions which remain are entirely my own.

Zelda Isaacson
September 1986

The author and publishers would like to thank the following for permission to use material in this book:
HMSO for the extracts from *The National Criteria – Mathematics* (Figures 1–26 of Chapter 1; Appendix B); The Southern Examining Group for examination questions on pp. 30, 31, 32, 33, 35; The Northern Examining Association for examination questions on pp. 32, 34, 35, 47; The London and East Anglian Group for examination questions on pp. 31, 33, 45, 46; The Midland Examining Group for the examination question on p. 36; the Shell Centre for Mathematical Education for Figures 4, 5, and 6 in Chapter 3; The Association of Teachers of Mathematics for Figure 8 in Chapter 3; the SMILE centre for Figures 5, 11(a) and 13 of Chapter 4 and Figures 1 and 2 of Chapter 5; the GAIM project at King's College London for Figure 9 of Chapter 4; Ward Lock for the extract on pp. 54–55 taken from Harvey, R., Kerslake, D., Shuard, H., and Torbe, B., *Language Teaching and Learning: 6 Mathematics*.

Introduction

In recent years there have been many exciting developments in mathematics teaching, many of them generated by classroom teachers. New ways of teaching mathematics have been tried out, often by teachers working within curriculum development projects, and have been very successful at improving pupils' motivation, confidence and standard of competence. Sometimes these have led to Mode 2 and Mode 3 examinations – that is, examinations designed by one or more schools for their pupils, usually internally assessed and externally moderated (or some variation of this pattern).

In many cases, change was motivated by the teachers' feelings about the effects on their pupils of standard Mode 1 examinations. One of the worst aspects of O level and CSE mathematics has been that conscientious teachers were forced to spend a lot of their time and energy (and that of their pupils) teaching how to pass mathematics examinations, instead of being able to concentrate on creating good conditions for learning mathematics. Syllabuses have been too large, often idiosyncratic, and with content which was conceptually beyond many of the pupils for whom it was intended. The pass-or-fail style of examination and the amount of necessary preparation put a strain on teachers and pupils alike. Many teachers have felt that they had to make an impossible decision – either (a) to put their emphasis on understanding, usefulness and pleasure, and 'cover' too little syllabus content for their pupils to be entered for an examination which properly reflected their ability relative to others, or (b) to teach towards an examination to the detriment of enjoyment and depth of understanding.

It is perhaps not surprising, then, that many young people leave school at 16 without the confidence to tackle even simple mathematical problems and with a deeply rooted fear of mathematics. Many adults suffer feelings of acute anxiety when confronted with the need to use mathematics at work or at home. Mathematics for most children (and adults) is a collection of meaningless symbols, inaccurately or hazily remembered rules and techniques, where answers are either right or wrong – and the judgement on whether an answer is right or wrong is always made by an authority, whether teacher or textbook. Thus judging for oneself whether an answer is sensible, appropriate, likely to be accurate or useful, is not possible when none of it makes sense anyway.

The experience of teachers and the evidence of research have influenced the proposals for GCSE examinations in such a way that they offer the possibility of a significant improvement. Most importantly, syllabus content will be substantially smaller, especially at the lower assessment levels. This should mean there will be time to teach for understanding, to consider a wide range of applications in realistic situations, to develop problem-solving skills, to do mathematics for fun and build on children's natural interest in puzzles and games, to enjoy the aesthetic aspects of mathematics, and to present work in attractive and appropriate ways. The qualities, abilities and knowledge which can then be developed will be accorded proper recognition when grades are awarded; awards will be made for positive achievement across a range of skills, concepts and knowledge at a level which is appropriate to the individual, and the inclusion of assessed coursework means that all will no longer depend on a candidate's performance on a single day in the artificial situation of a timed examination.

This book seeks to evaluate the proposed changes, pick out the similarities and differences between the existing examinations and the GCSE and focus (in order to offer practical guidance) on those elements of the GCSE which are 'really new' – in particular oral work, practical and investigative work and teacher assessment of coursework. The discussion of the latter aspects occurs mainly in Chapters 3, 4 and 5, while Chapters 1 and 2 deal with the national criteria and the new syllabuses. Readers who are familiar with the requirements of the new examination may prefer to start reading from Chapter 3, making use of the first two chapters and Appendix B for reference.

1 The National Criteria

> 'I only took the regular course,' said the Mock Turtle.
> 'What was that?' enquired Alice.
> 'Reeling and Writhing, of course, to begin with,' the Mock Turtle replied; 'and then the different branches of Arithmetic – Ambition, Distraction, Uglification and Derision.'
>
> *Lewis Carroll*

DEPARTMENT OF EDUCATION AND SCIENCE PUBLICATIONS

The key documents on the GCSE examinations in mathematics issued by the Department of Education and Science and the Welsh Office are:

GCSE: A General Introduction (DES, 1985a)
GCSE: The National Criteria – General Criteria (DES, 1985b)
GCSE: The National Criteria – Mathematics (DES, 1985c)

Most of this chapter consists of a close examination of the national criteria for mathematics, but to put this in context there is first a discussion of the sections in the other two documents with which mathematics teachers need to be familiar.

THE MAIN FEATURES OF THE GCSE EXAMINATIONS

The *General Introduction* to the GCSE describes the main features of the new examination system.

Fig. 1 ☐ It would be administered by five Groups of GCE and CSE Boards and monitored by the Secondary Examinations Council;

Figure 1: This is an enormous simplification of the existing system, and will almost certainly result in a much smaller number of syllabuses as well as a small number of bodies with whom schools and teachers have to liaise. All information from the constituent boards in a group is now available from any of the group's offices. This tidying-up of the current proliferation of boards and syllabuses is greatly to be welcomed. The right of individual schools (or groups of schools) to put forward their own syllabuses for consideration by an examining group has not been removed, and so the possibility of local (and national) curriculum development projects remains.

Fig. 2 ☐ all syllabuses and assessment and grading procedures would be based on the national criteria which had been prepared in draft by the Joint Council, as approved by the Secretaries of State for Education and Science and for Wales;

Figure 2: The Joint Council referred to here is the Joint Council of GCE and CSE Boards which has now become the Joint Council for GCSE. The National Criteria (general and subject-specific) have been approved, and all syllabuses for examination from 1988 onwards are based on these. This has resulted in much greater similarity among the syllabuses available than in the past.

Fig. 3 ☐ differentiated assessment techniques would be used in all subjects so as to enable all candidates to demonstrate what they know, understand and can do;

Figure 3: The intention that examinations should enable candidates to 'demonstrate what they know, understand and can do' is an extremely important and heartily to be welcomed feature of the GCSE. GCE O level examinations were certainly not devised with this intention. Built into the system was the expectation of failure. Not only were these examinations targeted at a very small proportion of the population (about the top 20 per cent) but also failure for approximately one-third of entrants was the *intention* of O level examinations. This partly arose because O levels were developed from the old School Certificate and Matriculation examinations, which were mainly concerned with eligibility for university entrance. As time went on, and these examinations were used more and more by employers in making decisions about taking on applicants for jobs, they acquired a general importance for the population as a whole which was not originally intended.

With the raising of the school leaving age to 16 in the 1970s, the need grew for an end-of-schooling examination for pupils who would not previously have been examined formally. Numbers entering for CSE examinations – always seen as a second-rate qualification hanging onto the coat-tails of O level – rose enormously. The CSE was bedevilled by the need to imitate O level (in order to gain respectability), although some teachers in some subject areas did their utmost to resist this tendency.

The opportunity afforded by the GCSE is, for the first time in this country, to develop a system of examining which is based on a 'bottom-up' approach. That is, it aims to start from the core of qualities, knowledge, skills and understanding which it is desirable for all future citizens to have acquired by the end of their formal schooling, and then to 'add on' to this core additional desirable qualities, knowledge etc., which only a proportion of the population may be able to obtain.

It remains to be seen, of course, what effect this statement of intent and the new syllabuses which are being offered will have in practice (in classrooms and in people's daily lives). Statements of intent, as is well known, do not always correlate well with outcomes. A CSE grade 1 pass is supposed to be equivalent to an O level pass *for all purposes*, but the reality is somewhat different! Some employers have even preferred to take a candidate with grade D or E at O level rather than one with CSE grade 1.

Nonetheless, the basic intention is, at the very least, a step in the right direction. Within mathematics it is to be put into practice by, amongst other things, having syllabuses with a smaller curriculum content (except those for the most able candidates) and requiring candidates to perform well over this smaller range of content rather than badly over a larger amount of content. Differentiated assessment in mathematics is to be achieved by having a series of examination papers of increasing scope and difficulty, so that all pupils sit examinations at a level of difficulty appropriate to their attainment. This is discussed in greater detail elsewhere in this book.

Fig. 4
☐ grades would be awarded on a single, 7-point scale, A, B, C, D, E, F and G: the standards expected of candidates would be no less exacting than in the existing examinations; the GCE Boards would bear special responsibility within the Groups for maintaining the standards of grades A to C and the CSE Boards would bear a similar responsibility for grades D to G;

Figure 4: This, regrettably, is to some extent an instance of the falling between two stools which is so difficult to avoid when trying to set up something new while retaining important features of the existing system. It would have been preferable, from many points of view, to make a completely fresh start rather than link new grades with old. However, in practical terms it is necessary to maintain some comparability between old and new systems of examining. Many employers, for example, are going to find it difficult enough to get to grips with the new system. This would have been made impossibly difficult for them if they were expected to understand a totally new system with no means of making comparisons between old and new. And it is of course entirely reasonable to vest the responsibility for maintaining comparability in the bodies which have amassed enormous experience over many years of examining, although it seems a pity that CSE boards will no longer have any responsibility for what was previously their top grade. The disappointment is also considerably mitigated by the final feature (Figure 5).

Fig. 5
☐ criteria-related grades would be introduced as soon as practicable: these grades would be awarded to all candidates who reached the required standards and not to any pre-ordained proportion of candidates.

Figure 5: Initially, GCSE will be targeted at the same 60 per cent of the population as O level and CSE taken together, an inevitable outcome of the feature discussed above. However, this statement indicates the intention to move from a norm-referenced grading system to a criterion-reference grading system. (There is a discussion of the characteristics and essential differences of these below.) In the proposed new grading system, if standards of achievement rise, a greater proportion of the population may achieve GCSE passes.

Definitions of the terms employed are provided in the glossary of terms in the *General Criteria* (HMSO, 1985b). The key phrases are reproduced as Figures 6 and 7. These are technical definitions, and the examples which follow are intended to illustrate them. Probably the most important thing to bear in mind is that pure versions of either system of grading are very hard to find – hence the peculiarity of the illustrative examples!

Norm referencing

Norm-referenced grading is intended to describe a system under which grades are allocated to predetermined proportions of the entry for the subject concerned.

Fig. 6

Figure 6: At its baldest and crudest, norm referencing occurs when it is decided in advance what proportion of entrants for an examination will be deemed to have passed or gained a particular grade. The candidates are then placed in an order of merit, and the grades awarded. Suppose, for example, that in country X potential entrants to the civil service take an entrance examination. Two hundred candidates enter, and although in fact most would be suitable for the work, there are a limited number of posts and only a proportion of the candidates can be taken on. Suppose there are altogether 60 posts, 10 of which are at a superior grade (and salary). A merit order must be agreed on, and the top 10 candidates in the list are then offered the superior-grade posts, while the next 50 candidates get ordinary posts. Everyone else is unsuccessful, regardless of how well they have performed in the examination. An election could be seen as an extreme case of this system. Only the candidate(s) with the most votes are elected, regardless of how suitable everyone else might be.

O level and A level examinations, it is often said, use this sort of grading system, and as a result only a predetermined proportion of people entering will pass, regardless of their actual attainment. This is in fact only partly true, because of the activities of GCE boards in attempting to maintain comparability of standards. In other words, an element of criteria referencing affects the outcomes, and the cut-off point between success and failure is determined to some extent by the boards' experience of what a pass (or particular grade) in a subject consists of.

Criteria referencing

Criteria-referenced grading is intended to describe a system under which grades are defined and awarded in terms of predetermined standards of performance specific to the subject concerned.

Fig. 7

Figure 7: This describes a system where success depends upon having reached a particular and definable standard (or where decisions on action are taken based on some objective criteria). The driving test is often quoted as an example of a test which employs criteria referencing. Whether or not you pass your driving test should depend on whether you are a safe and adequate driver, according to well-defined and objective standards and not on how well you drive in relation to other people. Another example is of medical decisions – such as whether a child needs to have his or her tonsils removed. This should depend entirely on objective criteria.

However, many people are sceptical about objective assessment and about criteria referencing in examinations. As regards the objectivity of the driving test, for example, people believe that whether or not one passes can depend on, amongst other things, the quality of driving seen earlier that day by the examiner – and such purely subjective things as the examiner's mood. In other words, even supposedly objective, criteria-referenced decisions are influenced by a norm, whether this norm is made explicit or not. This can be a norm of health (as in the tonsils example) or of skills and knowledge (as in the driving test example). The norm is derived from being exposed to other cases with which a (sometimes unconscious) comparison is made, and judgements which are supposed to be based on purely objective criteria are affected.

The statement in the *General Criteria* (p. 22) that 'both GCE and CSE must be regarded as being in practice hybrids, involving elements of both approaches' is undoubtedly true. However, the implication which underlies statements in the official documentation, i.e. that it is possible to have a 'pure' criteria-referenced grading system (and that this is what the GCSE will have) is not justifiable. Completely objective assessment of anything is impossible. Tightening up criteria and having checks and safeguards can result in greater objectivity, but an element of subjectivity, and of comparisons with a norm, is unavoidable.

Mathematics for all and mathematics everywhere

19

(h) *Avoidance of bias*

Every possible effort must be made to ensure that syllabuses and examinations are free of political, ethnic, gender and other forms of bias.

(i) *Recognition of cultural diversity*

In devising syllabuses and setting question papers Examining Groups should bear in mind the linguistic and cultural diversity of society. The value to all candidates of incorporating material which reflects this diversity should be recognised.

Fig. 8

19

(j) *Language*

The language used in question papers (both rubrics and questions) must be clear, precise and intelligible to candidates throughout the range of entry for the examination. Examining Groups should consider whether they need to make special provision for candidates whose mother tongue is not English.

Fig. 9

Figure 8: There is a growing recognition that positive steps can be taken to eliminate, or at least reduce, bias in examinations. The underachievement of, for example, girls in mathematics, science and technology can be partly attributed to textbooks and examination papers which discriminate against them, by ignoring their existence and interests. The invisibility of girls and women in the classroom materials for these subjects is well documented, as is the tendency for applications to concentrate on traditional 'male' interests. Members of ethnic minority groups too are grossly under-represented in textbooks for all subjects, including mathematics, and it would be hard to believe, if one glances through many books, that we live in a culturally diverse and richly varied society.

These statements are therefore most welcome, and it is vital that teachers are fully aware of them. There are important implications for the day-to-day work of classrooms, as it would be a great pity if syllabuses and examinations were free of bias and welcoming of cultural diversity, but children's daily experiences were different. Also, it is important that teachers regularly scrutinize examination questions and bring any bias they notice to the attention of the examination groups.

Figure 9: Too often examinations have been couched in specialised and unnecessarily abstruse language. It must be the experience of many teachers that their pupils could cope perfectly adequately with the subject matter in a question but could not sort out the language in which that question was set. We have had to teach our pupils how to read this specialised language as an extra task, and this skill, once acquired, was of no use except in other examinations. Similarly, rubrics were often complicated and difficult to interpret – generally, one had to make sure that pupils knew in advance what the rubric would state, as it was unrealistic to expect most of them to make sense of it on the day. Use of clear, straightforward language will benefit everyone, but perhaps especially those who have had additional difficulties through working in a language which is not their mother tongue or those at the lower end of the attainment range.

Figure 10: This paragraph is making three important points. In the context of mathematics these are: first, mathematics has a very close relationship with other subject areas, especially science and technology. Second, it is a reminder of the fact that mathematics is to be found all around us and impinges significantly on our daily lives. And last, but not least, it points out that as teachers of mathematics we should be aware, and try to help our pupils become aware, of the many social, political, economic and environmental issues which may be related to mathematical activities.

> 19
>
> (k) *Emphases to be encouraged in all subjects*
>
> All syllabuses should be designed to help candidates to understand the subject's relationship to other areas of study and its relevance to the candidates' own life. Awareness of economic, political, social and environmental factors relevant to the subject should be encouraged wherever appropriate. Questions seeking to test this awareness should be in the context of the subject concerned and not be independent of it.

Fig. 10

THE NATIONAL CRITERIA FOR MATHEMATICS

The document on the national criteria for mathematics is very short and warrants close study. The discussion in this section focuses on the aims, assessment objectives and content lists set out there and looks at the relationship between them. Techniques of assessment and grade descriptions are considered in Chapter 5.

Aims and objectives

Readers may well be puzzled about the difference between aims and objectives – and with good reason, because until recently, in educational writing, the two terms were often used more or less interchangeably. Nowadays, the usage which has been generally adopted is to reserve the word *aims* for the broader underlying purposes of an activity or examination and to use *objectives* to refer to the more concrete, usually more easily measurable, intended outcomes. So we might suggest that an aim of health education is to encourage young people to be aware of the importance of good diet for health, while a related, and measurable, objective might be that pupils should be able to plan a healthy diet for a group of people on a particular income and in particular circumstances. A scheme of work in mathematics might have as its aim the development of spatial concepts, while the objective of a particular lesson might be that pupils should construct solid shapes of particular dimensions.

It is this usage of the two terms which is employed in the *National Criteria – Mathematics*, and therefore in syllabuses which are drawn up in accordance with these criteria. To emphasise this distinction, the section on objectives is headed 'Assessment Objectives'.

The next part of this chapter consists of explanatory notes to the lists of aims and assessment objectives. These sections in the

Fig. 11

2. Aims

The statement which follows sets out ideal educational aims for all those following courses in Mathematics which lead to GCSE examinations. Some of these aims refer to the development of attributes and qualities which it might not be possible, or desirable, to assess directly.

Aims

Fig. 12

All courses should enable pupils to:

2.1 develop their mathematical knowledge and oral, written and practical skills in a manner which encourages confidence;

Fig. 13

2.2 read mathematics, and write and talk about the subject in a variety of ways;

Fig. 14

2.3 develop a feel for number, carry out calculations and understand the significance of the results obtained;

Fig. 15

2.4 apply mathematics in everyday situations and develop an understanding of the part which mathematics plays in the world around them;

Fig. 16

2.5 solve problems, present the solutions clearly, check and interpret the results;

2.6 develop an understanding of mathematical principles;

Fig. 17

2.7 recognise when and how a situation may be represented mathematically, identify and interpret relevant factors and, where necessary, select an appropriate mathematical method to solve the problem;

document are reproduced in full. The purposed of doing this is to put what may appear strange and new into the context of aspects of mathematics education which are very familiar, and to pick out (for detailed discussion elsewhere in this book) those aspects which are really new. These are very few. Most of the changes involve a difference of approach or of emphasis, rather than the introduction of something novel.

Figure 11: This is a clear statement that not only might it not be *possible* to assess directly some of the attributes and qualities to be aimed for; it would also not be *desirable* to do this. Such a statement is worth remembering. We tend to believe that we ought only to be teaching those things that are going to be assessed, as otherwise we are doing our pupils a disservice. This is not true. Some things which cannot be assessed directly nevertheless influence learning in other ways, such as by improving interest, motivation, autonomy, confidence or self-esteem.

Figure 12: 2.1 Mathematical competence is needed in verbal interactions and in practical activities as well as in written work. Traditionally, written work has predominated, and the shift towards more oral work is a welcome one. (I admit there has been 'mental arithmetic' in the past, but this is not the same as talking about mathematics.) Oral and practical work appear as 'new' objectives in Section 3. There is an extensive discussion of these in Chapter 3.

Figure 13: 2.2 The key word here is *variety*. Also, the importance of talk in mathematics learning has been reiterated.

Figure 14: 2.3 A *feel* for number and understanding of significance, rather than mere technical skill, is the way numeracy is best understood in the age of the calculator. Understanding what numerical procedures are appropriate and whether an answer is sensible are far more important than the ability to do lengthy pencil-and-paper calculations. It no longer matters whether people can do long division 'by hand', but it matters very much that they can decide that long division is the appropriate calculation to do, and can then assess whether their answer is reasonable.

Figure 15: 2.4 This has, I think, always been an aim of mathematics education. It remains to be seen whether it will be more easily realised with the new syllabuses. In Chapter 3 a few quotations from past statements on mathematics education serve as a reminder of how difficult it has always been to realise this particular aim.

Figure 16: 2.5 and 2.6 are old, familiar aims of mathematics education. We have not, in the past, been terribly successful with these.

Figure 17: 2.7 has familiar elements in it, but the notion of pupils at this stage of their education being capable of making these kinds of judgement is fairly new. There is a hint of 'mathematical modelling' in this statement.

Fig. 18

2.8 use mathematics as a means of communication with emphasis on the use of clear expression;

Fig. 19

2.9 develop an ability to apply mathematics in other subjects, particularly science and technology;

Fig. 20

2.10 develop the abilities to reason logically, to classify, to generalise and to prove;

Fig. 21

2.11 appreciate patterns and relationships in mathematics;

Fig. 22

2.12 produce and appreciate imaginative and creative work arising from mathematical ideas;

Figure 18: 2.8 This too is a familiar aim, but one which is given rather more emphasis than in the past. Using mathematics as a means of communication, although this is not the same thing as talking and writing about mathematics (2.2), involves this and also understanding of mathematical principles (2.6).

Figure 19: 2.9 is another familiar aim which we have failed to achieve to an alarming degree and which has caused much anxiety. The number of times the physics department in a school has blamed the mathematics department for inadequate teaching because pupils are quite unable to use simple algebraic and graphical techniques for their physics must be enormous! Science teachers sadly (or angrily) complain that they have to teach their pupils mathematics before they can teach them physics. Although their distress is understandable, it is not entirely fair. There are real difficulties which arise partly from questions of timing, e.g. pupils may be required to work with formulae in physics at an earlier stage than is appropriate to their mathematical development. Also, transfer of skills from one subject area to another is not a straightforward matter. (One or two of the quotations in Chapter 3 bear witness that this has long been a problem.) These difficulties point to the need for teachers to make explicit the cross-curricular links, and for more communication and consultation between teachers of different subject areas.

Figure 20: 2.10, again, is not new, but is given greater emphasis. There is nowadays a greater realisation that a deeper appreciation of the importance of classification, generalisation and proof, and a better understanding of the meaning of proof, come much more from engaging in these activities at one's own level than from merely rote-learning someone else's tidied-up work. Hence the importance nowadays attached to exploratory and investigative activities – but this is covered in greater detail later.

Figure 21: 2.11 Appreciating patterns and relationships in mathematics is also – or ought to have been – an old, familiar aim, but one which has not, I think, had sufficient emphasis placed on its realisation. This ability is the cornerstone of mathematical understanding and development and cannot be overemphasised. Much investigational work also depends upon – and helps develop – this appreciation. Many people, when asked for a definition of mathematics, say that mathematics is the study of relationships. This definition is somewhat more helpful than that old favourite, 'mathematics is what is written in mathematics books' or, 'mathematics is what mathematicians do'!

Figure 22: 2.12 At last – something in the aims that looks really new. The notion that mathematics for all (rather than mathematics for a gifted few only) can and ought to have an aesthetic component, and feed and help develop our creative and imaginative faculties is much to be welcomed. We need to combat the belief that mathematics is dry, cold and just about logic, by emphasising that mathematics is also about beauty, elegance, excitement and imagination.

Figure 23: 2.13 This is another new-looking aim. It expresses the view that mathematical activity can be practical and can involve cooperation with others. Some people in mathematics education have for a long time been expressing the view that extended pieces of work, experiment, and that magic phrase 'work of an investigative kind', are appropriate activities for the classroom. Here this view is given legitimation in a formal document.

Fig. 23

2.13 develop their mathematical abilites by considering problems and conducting individual and cooperative enquiry and experiment, including extended pieces of work of a practical and investigative kind;

These two 'new-looking' aims were of course laid out and extensively discussed in both the Cockcroft Report (DES, 1982) and *Mathematics from 5 to 16* (DES, 1985d). They are new only in the sense that most mathematics curricula have not included them.

Figure 24: 2.14 and 2.15 There is nothing new in either of these. In fact, the mathematics curriculum in the past has been rather distorted for the many by the needs of the few who were going on to study mathematics at higher education level. The requirements of the universities, demanding a particular content at A level, which then influenced the content at O level (and CSE), have had a largely negative effect on most children's learning of mathematics.

Fig. 24

2.14 appreciate the interdependence of different branches of mathematics;

2.15 acquire a foundation appropriate to their further study of mathematics and of other disciplines

To summarise, then, the 'really new' elements in these aims, from the point of view of everyday classroom activities are:

– the emphasis on oral work
– a shift in the meaning of numeracy arising at least in part out of the availability of cheap calculators
– mathematics for all pupils is to be understood as having a creative and imaginative dimension
– collaborative work, experimental and extended pieces of work, both practical and investigative, are legitimate forms of activity in school mathematics

Assessment objectives

It is worth stressing that objectives are the more concrete, more immediate and therefore more easily measurable outcomes of engaging in an activity. If one takes as an example an infant-school grouping activity, the aim might be to provide a foundation for the development of the concept of place value, while the objective might be that children should be able to group in threes and exchange a group of three for an object which represents the group.

Figure 25: This emphasis on assessment is underlined by the label 'Assessment objectives' and again in the introductory paragraph, reproduced here.

Fig. 25

3. Assessment objectives

The objectives which follow set out essential mathematical processes in which candidates' attainment will be assessed. They form a minimum list of qualities, abilities and skills. The weight attached to each of these objectives may vary for different levels of assessment within a differentiated system.

The assessment objectives constitute a list of 'qualities, abilities and skills' which the GCSE examinations must *test*. Note also that, although all candidates will be assessed on all of these, their relative importance may be varied in order to take account of different abilities.

The first fourteen objectives contain few surprises:

> 3.1 recall, apply and interpret mathematical knowledge in the context of everyday situations;
>
> 3.2 set out mathematical work, including the solution of problems, in a logical and clear form using appropriate symbols and terminology;
>
> 3.3 organise, interpret and present information accurately in written, tabular, graphical and diagrammatic forms;
>
> 3.4 perform calculations by suitable methods;
>
> 3.5 use an electronic calculator;
>
> 3.6 understand systems of measurement in everyday use and make use of them in the solution of problems;
>
> 3.7 estimate, approximate and work to degrees of accuracy appropriate to the context;
>
> 3.8 use mathematical and other instruments to measure and to draw to an acceptable degree of accuracy;
>
> 3.9 recognise patterns and structures in a variety of situations, and form generalisations;
>
> 3.10 interpret, transform and make appropriate use of mathematical statements expressed in words or symbols;
>
> 3.11 recognise and use spatial relationships in two and three dimensions, particularly in solving problems;
>
> 3.12 analyse a problem, select a suitable strategy and apply an appropriate technique to obtain its solution;
>
> 3.13 apply combinations of mathematical skills and techniques in problem solving;
>
> 3.14 make logical deductions from given mathematical data.

Comments: 3.4 and 3.5 are appropriate responses to the wide availability of calculators. Learning to use a calculator effectively is much more important than being able to remember and reproduce accurately a complex algorithm. These objectives were wholly predictable in the light of aim 2.3.

3.7 This is a familiar and very important objective, the attainment of which can be encouraged through sensible use of a calculator.

3.9 Although not a new objective, it is perhaps made more explicit than we are used to. This relates, in part, to aims 2.10 and 2.11 – so my comments there, on the value and importance of investigational work, apply here.

Only three objectives remain to be considered.

> 3.15 respond to a problem relating to a relatively unstructured situation by translating it into an appropriately structured form.

This is (relatively) new and very welcome. For too long mathematics education has failed children by presenting mathematics to them as neat and cut-and-dried. Real problems are not like that. Part of the skill of a mathematician is to take an unstructured situation and impose structure on it – and then see

Fig. 26
Two further assessment objectives can be fully realised only by assessing work carried out by candidates in addition to time-limited written examinations. From 1988 to 1990 all Examining Groups must provide at least one scheme which includes some elements of these two objectives. From 1991 these objectives must be realised fully in all schemes.

what results from that structure, and, perhaps, whether a different structure would be more helpful or illuminating for the problem.

The paragraph between 3.15 and 3.16 is very important (Figure 26). The examining groups must, from the first GCSE examinations in 1988, provide a coursework option, but this is not a requirement for mathematics until 1991. This means that pupils in their *first* year of secondary schooling in 1986–7 are the *first group* for whom the coursework element will be compulsory.

For many children, then, the last two objectives will not be fully assessed until 1991. These are:

> 3.16 respond orally to questions about mathematics, discuss mathematical ideas and carry out mental calculations.

Objective 3.16 is causing a lot of anxiety. 'Mental arithmetic' is not new, but has not traditionally been tested in O level and CSE examinations. Being assessed on one's ability to talk about mathematics is new. (A substantial section of Chapter 3 is devoted to oral work in mathematics.) Some of the new syllabuses have 'aural tests', that is, tests in which pupils have to listen to a question read by the teacher, usually carry out mental calculations (or derive information from a table) and write down their response. In some respects these are akin to the mental arithmetic tests of the past. Aural tests are usually in addition to 'oral assessment' which involves talking about mathematical ideas.

And finally:

> 3.17 carry out practical and investigational work, and undertake extended pieces of work.

This is perhaps the 'newest' objective of all, and like 3.16, is causing much anxiety. This too is fully addressed in Chapter 3 and also in Chapter 4.

The assessment objectives which are relatively unfamiliar, then, are:

– use of an electronic calculator (3.5)
– working with a relatively unstructured situation and imposing structure on it (3.15)
– oral work in mathematics; talk about mathematics and mental calculations (3.16)
– extended pieces of work: practical and investigative work (3.17).

Compare this list with the list of 'new' aims given earlier. There is, as one would expect, a very close correlation between them. Note however that one of the unfamiliar aims (2.12, the imaginative and creative aim) does not have an exact correlate amongst the objectives – but remember that it is not necessary (or appropriate) to assess *all* aims. Also, objective 3.15 relates to familiar aims (2.10 and 2.11). What is new, as was stated above, is the recognition that these aims are better realised by the pupil 'being a mathematician', at however elementary a level, than by having it all organised and tidied up. So there is, as one would expect, a close but not exact match between the new aims and the new-looking objectives.

Content

The content lists in the *National Criteria* are the closest we in England and Wales have ever come to mathematical syllabuses being prescribed nationally. (See Appendix B.) You will see there that the national criteria require that there should be at least three levels of assessment 'in any differentiated scheme offering the full range of grades'. This is a reminder that one of the central features of the GCSE is that 'differentiated assessment techniques must be used so as to enable all pupils to demonstrate what they know, understand and can do'. The major technique for differentiation in the assessment of mathematics is in fact the use of several levels of examination with different syllabuses and examination papers.

There are two lists. The content in List 1 is intended to be included in the curriculum of all pupils and thus constitutes a common core of curriculum content. List 1 is also to be almost all of the curriculum content for the lowest-level candidates. List 2 is most of the additional content which is to be required for the middle assessment level.

There is a very close resemblance between List 1 and the foundation list in the Cockcroft Report, paragraph 458 (DES, 1982). Cockcroft's foundation list and both the lists in the national criteria have been developed as a result of a great deal of discussion and consultation within the mathematics education community. They have been guided by research evidence which has given us information about: how children learn mathematics; which areas of the mathematics curriculum create the greatest difficulties for learners; and the mathematics knowledge and skills which are most needed in adult life.

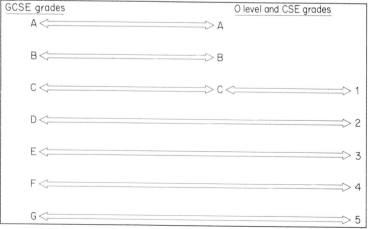

▲ *Fig. 27*

Another important feature of GCSE which is reflected in this section on content is the statement that grades are to be awarded on a single seven-point scale, A to G, which relate to existing O level and CSE grades. Figure 27 illustrates the relationship between the old and new grading systems. Paragraph 523 of the Cockcroft Report also suggests 'seven grades of success' and relates these to O level and CSE grades in precisely this way.

The contents section states that the lowest-level examination, for which List 1 is to be most of the syllabus content, should be one in which the majority of candidates are expected to get grades E, F and G. The middle level should be aimed at grades C, D and E, and the highest level should be targeted at the most able candidates who aspire to grades A or B. The latter would cover a syllabus 'well in excess of that contained in List 1 and List 2'.

This is very much in accord with the views expressed in the Cockcroft Report.

> We consider that the content of the examination syllabus for those pupils who at present achieve around CSE grade 4 should not be very much greater than that of the foundation list ... At a higher level, we consider that a syllabus whose extent is comparable to that of existing O level syllabuses represents a suitable examination target for pupils in the top 20% of the range of attainment in mathematics. We believe that there is also need for an examination syllabus whose content lies between that of the two reference levels we have indicated ...
>
> (DES, 1982; para. 472)

The way in which the content requirements at each level are 'nested' inside each other reinforces the notion of a common core. The 'core' items, in Lists 1 and 2, will not only be common from syllabus to syllabus around the country, but will also constitute a large part of the work of even the 'high flyers'. Any scheme of assessment for these candidates must allocate between 50 and 70 per cent of the marks carried by written examinations to questions which relate to List 1 and List 2 items.

One outcome of this will inevitably (and thankfully) be an enormous simplification of the existing system. Syllabuses provided by different boards and in different parts of the country will be much more like each other than in the past. This will make it much easier for both pupils and teachers to move from school to school or from one part of the country to another. The decision about which level a pupil should take can be reasonably delayed until quite late, as what is common will be greater than any differences. Coursework too, once this is included, will help to make the system more flexible, as candidates will be able to tackle coursework tasks at their own level. Candidates who do particularly well (or particularly badly) at the coursework elements could in some circumstances have their work considered for a different level and graded accordingly.

There is more detail about what this all means in practice in Chapter 2, which looks at how the examining groups are planning to put the *National Criteria* – general, and specific to mathematics – into effect.

One very important aspect to notice is that List 1 and List 2 together constitute a much smaller syllabus that we have been accustomed to at CSE level. And of course List 1 on its own, for candidates who in the past would have taken CSE examinations and obtained the lowest grades, is very much smaller than current CSE syllabuses. One of the very welcome outcomes of GCSE planning is an overall reduction in syllabus sizes, which is particularly marked at the lower levels. For most pupils there will be less syllabus content to 'cover' and therefore there will be more time to help our pupils develop their understanding of mathematics, for them to make the mathematics they are learning their own, and to relate their mathematical knowledge to their everyday needs.

When you are wondering how you are going to 'fit in' the new objectives, remember that much of the time which is currently spent on teaching inappropriate and too difficult content will be saved and will be available for making mathematics fun, relevant and satisfying.

Relationship between assessment objectives and content

This is a very small section (only two paragraphs) which does little more than state that there is not an exact match between particular objectives and particular items of content. It also states that at different levels of assessment different emphases would be appropriate.

2 The Examining Bodies and the New Syllabuses

> 'When I use a word,' Humpty Dumpty said, in rather a scornful tone, 'it means just what I choose it to mean – neither more nor less'.
>
> *Lewis Carroll*

REQUIREMENTS

This chapter examines the responses of the examining bodies to the requirements and criteria of the GCSE in mathematics as laid down in the official documents. Some of these requirements apply to all subjects; others are specific to mathematics. Some requirements are technical in nature, and need not concern us in this book; others concern teachers in their day-to-day work in the classroom. In the case of some of the requirements, the examining body's response can be understood through a reading of their published syllabuses, by looking at aims and objectives, schemes of assessment, lists of content, and so on. However, in other cases the ways in which the examining bodies are seeking to realise the national criteria in practice can only be properly understood by studying the specimen papers they have provided in order to see the style of questions they intend to set. Equally important are the variations in the coursework requirements of the groups. All these issues are addressed in this chapter.

What becomes apparent is that, as one would expect from the survey of the national criteria provided in Chapter 1, there is much that is familiar in GCSE mathematics. The changes are ones which will greatly improve pupils' mathematical experience. Many changes are ones of simplification and clarification. Others result from the recognition of important aspects of life in the late twentieth century or from acceptance of important technical changes.

Summary of requirements

The following is a summary of the requirements of the examining bodies with which all mathematics teachers need to be familiar:

1. All the existing O level and CSE examination boards are to be grouped together into *five examining groups*.
2. *Differentiated assessment techniques must be used* – to enable all candidates to demonstrate what they know, understand and can do.
3. *Grading must be on a seven-point scale* and must be related to standards in existing O level and CSE gradings.

4. *Criteria-related grades are to be introduced* and incorporated into the subject-specific national criteria as soon as practicable. At the time of writing this is expected to be 1991 or later.

5. a) *Syllabuses must be approved by the Secondary Examinations Council* and must comply with the aims and assessment objectives as laid down in the subject-specific criteria.
   b) *Mathematics syllabuses must provide a means of assessing objectives 3.16 and 3.17* (see Chapter 1). This coursework element, which includes the assessment of oral work, must be made available by each of the groups from the first GCSE examination, i.e. from 1988, but will not be compulsory until 1991.
   c) *Syllabuses must, similarly, comply with the requirements on content* as laid down in these criteria.
   d) *Syllabuses must comply with paragraphs 19(h)–(k)*. These are the paragraphs which concern avoidance of bias; recognition of cultural diversity; use of clear and precise language, which must be intelligible to candidates throughout the range of entry for the examination; and relevance to candidates' own lives. There is a discussion of these in Chapter 1.

THE EXAMINING GROUPS

The twenty-two different examination boards have been grouped into five larger examining bodies which a corresponding simplification of the 16+ examination system. The Groups are often referred to by their initials:

NEA   Northern Examining Association
MEG   Midland Examining Group
LEAG   London and East Anglian Group
SEG   Southern Examining Group
WJEC   Welsh Joint Education Committee

Some examining bodies, such as the boards which now constitute LEAG and those which constitute NEA, anticipated the development of a unified system of examining at 16+ by developing joint O level/CSE examinations. These were offered alongside conventional O level and CSE examinations, resulting in an even greater proliferation of syllabuses. This did, however, provide a useful (although not always positive) experience for the boards and the schools which participated.

DIFFERENTIATED ASSESSMENT TECHNIQUES

What does this mean for mathematics and how have the groups responded? In mathematics, differentiated assessment techniques fundamentally mean that there are several levels of assessment and that syllabuses of different size and accompanied by examination papers of different levels of difficulty are offered. Coursework tasks provide an additional means of differentiation.

Throughout the official documentation there is a clear requirement that differentiated assessment techniques are to be employed. Paragraph 16 of the *General Criteria* is the key statement on this. The primary reason for this requirement lies in the phrase, reiterated in several places in the documents, that all candidates must be given opportunities to show what they know, understand and can do. Examinations which are either too easy or too difficult do not provide these opportunities. It is emphasised that the lower grades must be awarded to candidates who are able to do well in an examination testing a limited range of skills or knowledge, rather than, as at present, by doing badly in an examination which is too difficult. This was in fact one of the recommendations of the Cockcroft Report.

A very useful document which lays out clearly the different techniques which can be employed to achieve differentiation is the SEC *Working Paper: Differentiated Assessment in GCSE* (SEC, 1985a). In addition to any other techniques which might be used, some subject areas, mathematics being one of these, must provide differentiated papers appropriate to different levels of assessment. (In other subjects differentiation can more easily by achieved by alternative methods.)

The mathematics criteria (paragraph 4.1 – reproduced in Appendix B) specify that *at least* three levels of assessment must be provided. Most of the new syllabuses do in fact have three levels. These are called variously, levels 1, 2 and 3 or levels P, Q and R, or foundation, intermediate and higher. There are all sorts of labelling systems around and these are included in the table (Figure 1) which summarises key information from the groups. However, having three levels of assessment does not necessarily mean that three distinctively different sets of examination papers must be set. A very popular pattern is to offer four papers of increasing scope and difficulty. Candidates for the lowest level (foundation, say) would sit papers 1 and 2, candidates for the intermediate level would sit papers 2 and 3 and the highest-level papers would be 3 and 4. This is often called the 'four-in-line' model and is represented by Figure 2.

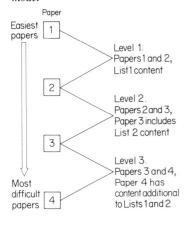

▼ *Fig. 2 The four-in-line model*

Out of the five examining groups, four use the 'four-in-line' model, and only one, the Midland Examining Group, intends to provide distinct pairs of papers for each of the three levels for its 'standard' syllabus. This is the 'three pairs' model; see Figure 3.

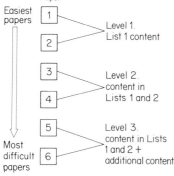

▼ *Fig. 3 The three-pairs model*

All the groups allow for some overlap in grading between levels. For example, the Southern Examining Group's arrangements, which are typical, are summarised in Figure 4.

A target grade has been specified for each level, and SEG's stated intention is that *the target grade is awarded for approximately two-thirds of the marks available* – because, in line with the national criteria, the assessment pattern has been designed to reward positive achievement. Thus the average pupil at any level would gain about 66 per cent. One grade below the target grade would be awarded for about half the marks available, so a pupil entered at an appropriate level should gain at least half the marks available – a welcome change indeed.

▼ *Fig. 4 Assessment pattern: Southern Examining Group*

Level	Grades available	Target Grade
1	E,F,G	F
2	C,D,E,F	D
3	A,B,C,D	B

Fig. 1 *Profiles of the GCSE examining groups*

	LEAG	MEG	NEA[1]	SEG	WJEC
Names of levels (from lowest to highest) and grades available at each level	*Level X*: E,F,G. Target grade F *Level Y*: C,D,E,F. Target grade D *Level Z*: A,B,C,D. Target grade B	*Foundation*: E,F,G. Intended target group – grades E,F,G. *Intermediate*: C,D,E,F. intended, target group – grades C,D,E. *Higher*: A,B,C,D. Intended target group – grades A,B,C.	*Level P*: E,F,G. Target grade not specified *Level Q*: C,D,E,F,G. Target grade not specified *Level R*: A,B,C,D,E,F,G. Target grade not specified	*Level 1*: E,F,G. Target grade F *Level 2*: C,D,E,F. Target grade D *Level 3*: A,B,C,D. Target grade B	*Level 1*: E,F,G. Target Grade F *Level 2*: C,D,E,F. Target grade D *Level 3*: A,B,C,D. Target Grade B
Proportion of total marks allocated to coursework and oral assessment (where offered)	25% for five tasks in three categories: oral assessment not specifically included	25% for five tasks, one from each of five categories: oral assessment may be included	25% subdivided as: practical investigational work 30 *assimilation 40 30 100 *includes 7 for 'oral communication'	40% for three units of work [one-fifth of this for oral assessment]	26% [70 out of 270 marks] for two tasks, [30 marks for each task, 9 of these for oral assessment]
Proportion of total marks allocated to aural tests	5% for 'mental' test, taken by coursework candidates only	None	None	10% for two aural tests taken by all candidates	10 out of the 70 coursework marks for an aural test based on the practical investigation

[1] Ranges of grades proposed by NEA not yet approved by Secondary Examinations Council

In the existing system, at CSE for example, where examiners are required to differentiate across five pass grades, grade 5 is awarded for very low marks indeed – in other words for failure at most of the examination. Differentiated assessment with at least three levels of assessment, together with limited grade examinations, are all essential if most of our pupils are to be enabled to perform well at an appropriate level.

However, the other side of this very positive coin is that the onus is on teachers to enter pupils for the appropriate level of assessment. If pupils are entered for a level which is too difficult, they will not simply get a poorer grade – they may get no grade at all. Similarly, a level which is too easy will not enable pupils to demonstrate their attainment appropriately. The first of these, that is, not entering pupils for a too difficult level of assessment, is perhaps, for many teachers, more of an adjustment than the second. On the other hand, teachers under pressure from parents to enter pupils at inappropriately high levels will now be better able to dissuade them by pointing out the risk which this involves. Parents who are anxious that their children should get a grade equivalent to an O level pass might be reassured to know that this can be gained at the middle level of assessment.

The amount of overlap in the grades available at different levels will make the decisions easier than perhaps seems likely at first sight. There is also a large amount of overlap between the levels in the examinations set and in coursework topics, as well as in syllabus content. Teachers who are using a syllabus with a coursework element will also acquire valuable information on their pupils' level of attainment when the coursework tasks are assessed, which will be prior to the examinations.

It may be helpful to remember that, in the four-in-a-line model, about two-thirds of all candidates will sit Paper 2, and similarly about two-thirds will sit Paper 3. (This assumes that approximately equal numbers of candidates are entered for each level.) Also the content at each level is included in the content at the next level, and so on. Even at the highest level of assessment, questions on the core content in Lists 1 and 2 are to account for between 50 and 70 per cent of the available marks.

There is still a certain amount of individual freedom – schools, or groups of schools, can as before submit their own syllabuses for validation by one of the groups. However, the examination groups must ensure that these Mode 2 or Mode 3 syllabuses comply with the national criteria. The Secondary Examinations Council is charged with monitoring the syllabuses. So the pattern of the syllabus that your school is planning to use may be different from any of the standard Mode 1 syllabuses provided by the examining groups, but will have important elements in common with all other mathematics syllabuses.

There is also greater freedom over the choice of syllabus. In the old system, schools had to enter their CSE candidates for a syllabus offered by their regional boards. In the new system, any school may use a syllabus offered by any of the groups for candidates at all levels.

A CHECKLIST FOR SCRUTINISING A MATHEMATICS SYLLABUS

You might need to scrutinise syllabuses prior to selecting one for your pupils, or, having already made this decision, you might need close knowledge of the syllabus in order to prepare a teaching scheme. In either event, it is useful to have a checklist and make notes. You can then very easily compare one syllabus with another, weigh up pros and cons for your particular situation and make informed decisions. Or you can compare the GCSE syllabus you have chosen with the O level and CSE syllabuses you have been using in order to see clearly where changes need to be made to your schemes of work. What follows is a suggestion for such a checklist.

A major factor, and one which will greatly influence the associated teaching schemes, is whether or not your syllabus includes coursework – that is, school-based or centre-based assessment – at this stage.

This is perhaps the appropriate place to reiterate that coursework in *mathematics* is optional for 1988, 1989 and 1990. All the groups are therefore offering a choice between syllabuses (or schemes within the same syllabus) where one version includes coursework and the other does not. Any school which lacks experience of coursework in mathematics would be well advised to take things gently and for the time being enter pupils for syllabuses which do *not* include this. There is time in the interim for both teachers and pupils to try out new ways of working, such as investigational and practical work. Pupils who enter secondary school in 1986–7 are the first for whom these elements will be compulsory, and your experimentation could be carried out with them, well before they reach the public examination stage.

You will need highlighting pens, an approved mathematics syllabus and specimen papers. You will also need to refer to Chapter 1 of this book (for lists of aims and assessment objectives from *National Criteria – Mathematics* and Appendix B (for content lists).

Aims

These will probably be identical to the aims in the *National Criteria – Mathematics*. If not, differences will usually be slight. Check by using a highlighting pen to mark the documentation from the examining body. Note any differences on a sheet of paper.

Objectives

Do the same as above, comparing with the assessment objectives. A syllabus *without* school-based assessment will almost certainly omit 3.16 and 3.17. These are the objectives 'which can be fully realised only by assessing work carried out ... in addition to time-limited written examinations'.

Content

Do the same as above (i.e. marking the syllabus with a highlighting pen) for each of the three levels. This will be more difficult, as content tends to be differently organised, often with topic headings. The easy way to do this and be sure you have not missed anything is to mark the syllabus, as before, with a highlighting pen and underline corresponding phrases in Lists 1 and 2 in soft pencil.

Check the lowest level against List 1. Note any differences. Check the middle level against Lists 1 and 2. Note any differences. For the highest level, list the additional content (i.e. over and above Lists 1 and 2) and note any other differences.

Read the notes on the content supplied by the group to get a feel for what is actually required.

Assessment

Look at the scheme of assessment and find out:

Timed papers

1. Which model has been adopted (e.g. four-in-line, three pairs)?
2. Is there choice on any of the papers?
3. Are there any multiple-choice sections or papers?
4. What proportion of the total marks is allocated to each paper?

Oral work

1. Is there an oral test? How is this organised?
2. What proportion of total marks is allocated to this?
3. Is oral assessment explicitly included in the coursework?

Coursework

1. What proportion of the total is allocated to this?
2. How many tasks are there? What sort of tasks?
3. How are tasks selected? Are there any constraints?
4. Do schools or individual pupils have any choice?
5. Do groups offer tasks, and in how much detail?
6. How is the assessment carried out – who does it, and who moderates?

General

Are any formulae sheets provided for the timed papers? Are there any recommendation on calculators? Are other calculating aids or tables allowed?

Comments

An outline of the coursework requirements for each of the groups is provided at the end of this chapter, but you will need more detail, particularly with regard to what the groups actually mean by a task with a particular label ('investigations', for example, mean different things to different people) and the amount of guidance offered.

Once you have done this, for several syllabuses if necessary, you will have a feel for what is required. This is a task which could well be shared among several colleagues, especially if you do have to scrutinise several syllabuses prior to selection.

The next stage, when you have perhaps narrowed the choice down to two syllabuses or have chosen one, is to read and work through the specimen papers the boards provide, and look at the mark schemes or comments on marking. The style of questioning adopted by a board is an aspect of your final choice, and for this only looking at specimen questions will help. A few examples are provided later in this chapter. It is here that one can see whether groups have really adopted the anti-sexist, anti-racist and multicultural recommendations, as well as the requirements concerning use of clear and comprehensible language and of relevance to everyday life and other subject areas. Many of the specimen questions are in fact taken from past O level and CSE papers.

Your final choice of syllabus for the next year or two will have been made after a careful consideration of your particular needs at this stage. There is no 'best' syllabus for all situations. The coursework now-or-later variable is clearly an important one, but so is the additional content required for the higher-level candidates. This is where the greatest variation in content among groups' syllabuses is to be found. It would be sensible to choose a syllabus where this additional content is familiar to you – and was perhaps included in your existing O level syllabus. This would minimise the need to prepare teaching materials for new content and to acquire new textbooks. Any available resources of time and money could be devoted to building up your equipment (human and material) for the new areas.

Another variable you might consider is whether or not a multiple-choice paper is included in the scheme of assessment. There is evidence that this sort of examination may discriminiate against girls, and this might deter you from using it in a girls' or mixed school. (In the new syllabuses the only group currently offering a multiple-choice paper is LEAG.) Finally, if you have opted for coursework, you may well prefer a group which offers specific tasks rather than just guidelines. On the other hand, you may like the freedom of choice which guidelines only offer. There are in fact many patterns available, and it is worth finding the pattern that best suits your needs.

The analysis you have now carried out will be the foundation on which you can build a teaching scheme. Ideally, choice of syllabus and construction of a teaching scheme should involve the mathematics staff as a whole. Everyone will have to operate it, and whatever is chosen is much more likely to be successful if everyone feels committed. The introduction of GCSE offers splendid opportunities for informal school-based inservice work. Ultimately, all GCSE mathematics teachers will be involved in assessment of coursework, and this will require collaborative work so that consensus can be reached on marking. One of the aims of GCSE is the encouragement of cooperative work (2.13), and this surely should apply to teachers as well as to pupils!

SYLLABUS CONTENT

Two key aspects of curriculum content in the new GCSE syllabuses are:

1. The relationship between the content prescribed by the groups and that laid down in the *National Criteria*.
2. The relationship between content prescribed for the different levels of assessment within a particular group's syllabus – that is, how progression and differentiation are to be achieved from one level to the next.

The first of these can be dealt with very quickly. Examining groups have followed the national criteria very closely. As indicated in the Checklist, the lowest level in each syllabus differs very little from List 1. The content for the middle level will be Lists 1 and 2 taken together, with minor additions, and the major variations between groups lie in the additional content each has prescribed for the higher-level examinations.

Progression between the levels is achieved in several ways:

– by 'adding on' further content to a topic already included;
– by including additional topics; and
– by extending content to allow for more difficult applications.

These aspects can be illustrated, and instances of progression seen, through an examination of content and specimen questions in particular topics.

A new style of question? – some specific examples

In this section, syllabus content and specimen questions relating to two topics, mensuration and application of number, are looked at across levels and examining groups.

As there is no consistent system for naming the levels, I shall simply refer to the lowest level as Level 1, the middle as Level 2 and the highest as Level 3, regardless of the actual names given to them by a particular group.

Mensuration

List 1 in the *National Criteria* specifies the following content:

– Perimeter and area of rectangle and triangle;
– Circumference of circle;
– Volume of cuboid.

List 2 adds the following:

– Area of parallelogram;
– Area of circle;
– Volume of a cylinder;
– Results of Pythagoras (if you regard this as a topic in mensuration).

All the groups have exactly the same as List 1 for their Level 1 content. At Level 2, all groups include List 1 and List 2 content, but some specify one or two more items, and some classify slightly differently. For example, LEAG classifies Pythagoras under trigonometry, NEA under mensuration, MEG has it as a separate item and SEG and WJEC follow the pattern in the *National Criteria* of simply giving a list without labelling subsections. In the LEAG, MEG and WJEC syllabuses, the area of a trapezium and the volume of a right prism are included as Level 2 content, while SEG adds nothing extra and NEA adds the area of the sector of a circle.

At Level 3 there are some variations. Most groups now include mensuration of the sphere, pyramid and cone; some restrict Pythagoras to two dimensions, as for Level 2; others

---

13. The pins on a pinboard are spaced 1 centimetre apart. The pinboard and an elastic band are used to show a variety of rectangles.

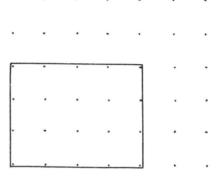

(a) What is the length of perimeter of the rectangle shown?

Answer .... *(2 marks)*

(b) On the same grid, draw another rectangle which has a perimeter equal to that of the rectangle shown, but with a different length and breadth.

(2 marks)

▲ *SEG, Level 1 (specimen question)*

extend this to three dimensions. There are other small differences, but nothing of great significance.

As was stated above, similarities between groups' syllabuses are very marked, and, predictably, at Levels 1 and 2 content follows the *National Criteria's* lists very closely.

The example questions are all identified by group and level. Where two levels are named, this is because the question either appears on a paper which is to be taken by candidates in both the levels, or because, as in the case of MEG which uses the 'three pairs' model, the same question appears in papers at two levels. The space provided for working on the question papers has been omitted.

What stands out when one scans specimen GCSE questions, and compares them with O level and CSE questions, is the familiarity of the curriculum content which is being tested, together with the genuine attempts of the groups (with varied success) to put this content into the context of everyday objects

▶ *LEAG, Level 1*

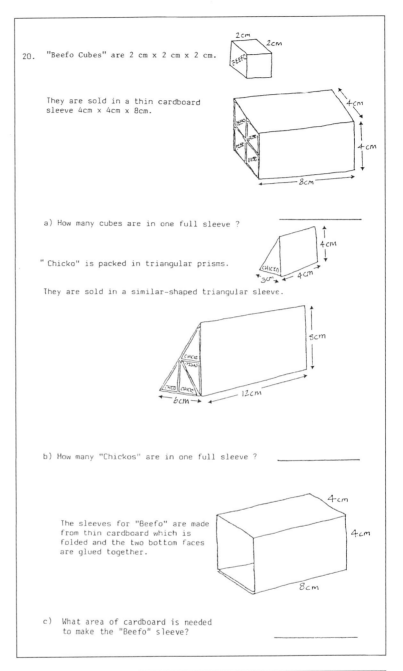

20. "Beefo Cubes" are 2 cm x 2 cm x 2 cm.

They are sold in a thin cardboard sleeve 4cm x 4cm x 8cm.

a) How many cubes are in one full sleeve? _____

"Chicko" is packed in triangular prisms.

They are sold in a similar-shaped triangular sleeve.

b) How many "Chickos" are in one full sleeve? _____

The sleeves for "Beefo" are made from thin cardboard which is folded and the two bottom faces are glued together.

c) What area of cardboard is needed to make the "Beefo" sleeve? _____

3. The instructions for erecting a greenhouse say: "First make a rectangular base 2 m by 3.5 m. To check that it is rectangular measure the diagonal. It should be about 4 m."

Using Pythagoras theorem, explain why the diagonal should be about 4 m.

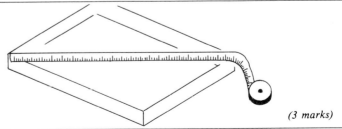

(3 marks)

▲ *SEG, Level 2/3 (specimen question)*

11.

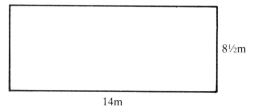

O is the centre of the circle.

The tables in a Burger Bar are circular, with a minor segment removed to form a straight edge. They have a diameter of 1 metre, and angle BOD is 90°. The tops are covered with formica and the perimeter is bound with thin steel strip. Calculate the area of the table top and the length of strip required. *(12 marks)*

▲ *SEG, Level 3 (specimen question)*

11 A man wants to put fertilizer on his lawn (shown here). His garden book tells him to use 50g of fertilizer for each square metre.

(i) How much fertilizer does he need? *(2)*

(ii) At the shop, he can get only one packet of fertilizer which weighs 5kg. He spreads this over the lawn. On average, how much fertilizer is there on each square metre of the lawn? *(3)*

▲ *NEA, Level 1/2*

6 Frojus are made of frozen fruit juice. They are sold in the shape of a triangular prism which is 11cm long and whose regular cross-section is an equilateral triangle of side 2.5cm.

Given that the fruit juice used increases in volume by 5% when it is frozen, calculate the volume of liquid fruit juice required to make one 'Froju'. *(5)*

▲ *NEA, Level 3*

and situations. The single most outstanding attribute of traditional mathematics papers has been that they are so often dull and arid. We can see, in some of these examples, a serious attempt to devise questions which make the purpose of the mathematics apparent – as indeed has been the case in some O level and CSE papers. For example, making burger bar tables and Frojus, and packing stock cubes, give significant context to problems of perimeter, area and volume. Not all the specimen questions are like this, however.

There is also, in some of the specimen questions (although not particularly in the mensuration examples), a human dimension – real people of both sexes, many races and all ages are faced with everyday choices, problems and tasks which require mathematical competence for their solution. Some striking examples of this solid and familiar reality occur, as one would expect and hope, in questions dealing with applications of number.

▲ *LEAG, Level 1*

Applications

In this area all the groups follow the national criteria very closely indeed. Personal and household finance, sterling and foreign currencies, applications of the 24-hour clock in reading timetables, the mathematical language of the media, reading of clocks and dials, profit and loss, etc. all appear at Level 1, with somewhat more difficult applications but little if any new content at Levels 2 and 3. It is in this area that a change of style can most easily be discerned, and the questions reproduced have been chosen to illustrate this. There are many more pictures which add interest and also test the ability of candidates to derive information from sources other than words alone.

9. Mr. and Mrs. Khan want to buy some "PLAIN BROADLOOM" carpet for their living room.

They need just under 20 square yards and want to pay less than £150.

Which of the PLAIN BROADLOOM carpets can they afford?

....

....

Carpets Galore

Probably the largest stock and choice in the UK, backed by a superb fitting staff...

A0 HEAVY DOMESTIC
27" BODY LOOM........................from £7.95 yd
12' BROADLOOM................from £13.95 sq yd

A00 LUXURY DOMESTIC
27" BODY................................from £11.95 yd
12' BROADLOOM................from £16.95 sq yd

PLAIN BROADLOOM
SQ. YD.
PREMIER £14.95
ROYAL HOUSEHOLD......................... £12.95
KING KURL TWIST £10.95
LAKELAND TWIST £9.95
COUNTRY COUSINS £9.95
ROYAL HARMONY £7.95
SHEARLING £6.99
PASTICHE £5.95
PRELUDE £4.95

▲ NEA, Level 1/2   ▼ SEG, Level 1/2 (specimen question)

15. The dials on the electricity meter at the home of Mrs Singh on December 31st are shown below.

1000   100   10   1

(a) What reading is shown by the dials?

Answer .... (4 marks)

On January 1st the meter is replaced by a digital meter. The meter shows a reading of 60 units when installed, as shown

| 0 | 0 | 6 | 0 |

(b) Between January 1st and March 31st Mrs Singh uses 742 units of electricity. Show on the meter below the reading on March 31st.

| | | | |

(4 marks)

(c) Mrs Singh receives a bill for the electricity used between January 1st and March 31st. She has to pay £7.28 (the fixed charge) plus 4.9p for every unit she uses.
Calculate the amount she has to pay.

Answer .... (8 marks)

(d) The next bill for the electricity used between April 1st and June 30th, is for £56.28. The fixed charge and the cost per unit have not changed. How many units of electricity has Mrs. Singh used this quarter?

Answer .... (8 marks)

14. The table below shows the charges made by the **TOPSKI HOLIDAY COMPANY** for skiing holidays to Austria. The prices shown are per person in **£**'s.

Date of departure	15 Dec	22 Dec	29 Dec	5 & 12 Jan	19 & 26 Jan	2 Feb	9 & 16 Feb	23 Feb	2 & 9 Mar	16 & 23 Mar	30 Mar
7 nights	174	210	284	215	184	204	226	235	204	190	198
14 nights	322	413	368	276	276	307	344	322	307	285	307
Reductions	Children aged 2 to 11 years: 50% reduction										

Use the table to answer the following questions.

(a) (i) Mrs. Jones plans to spend 7 nights in Austria. How much will it cost if she departs on 22nd December?

Answer .... *(1 mark)*

(ii) How much more would it cost if she departed one week later?

Answer .... *(3 marks)*

A family of two adults and one child aged 10 years, book a holiday to Austria for 14 nights. They depart on 16th February.

(b) (i) How much is the child's holiday? Answer .... *(2 marks)*

(ii) How much is the holiday for all three of them?

Answer .... *(2 marks)*

▲ *SEG, Level 1 (specimen question)*

▼ *LEAG, Level 2/3*

Violet intends to make herself an outfit to wear at her brother's wedding in October. She decides to make a pair of trousers, a blouse and a tunic. The table below shows the quantity of fabric required for her size and the cost per metre of the various materials she has chosen.

Item	Quantity in metres	Cost per metre
Trousers	2·40	£4.90
Blouse	2·90	£3.10
Tunic	1·50	£6.40
Lining	1·50	£1.40

a) Calculate the cost of each of these four items: (5)

b) Calculate the cost of

　i) 1 zip (18 cm) at 5p per cm,

　ii) 6 buttons at 9p each,

　iii) 3 reels of thread at 57p each,

　iv) 1.50 m of tape at 12p per metre. (4)

c) Calculate the TOTAL cost of the items listed in a) and b). (2)

d) Violet is entitled to a discount of 8% on the total cost.

　Calculate, to the nearest penny, how much she actually pays. (4)

ANSWERS

Cost of material for:

i) Trousers

£ _____

ii) Blouse

£ _____

iii) Tunic

£ _____

iv) Lining

£ _____

35

5. Melanie Crisp entered for a sponsored walk in aid of Oxfam. This is her sponsorship form. She walked 13 miles.

AMBRIDGE YOUTH CLUB: SPONSORED WALK for OXFAM: 31/8/85		
Signature of sponsor	Amount per mile	Amount given
Mrs Crisp	5p	
Harry Crisp.	1p	
Auntie Jan & Uncle Bill	25p	
B. Kay	3p	
Ben Johnson	15 pence	
J. E. Dooley	60 pence	
	GRAND TOTAL COLLECTED	

NAME OF WALKER ..... Melanie Crisp
Distance walked ..... 13 ..... Signed ..... B. T. Powell

(i) Complete the column 'Amount given' and fill in the box for the grand total that she collected. *(4)*

(ii) Melanie wanted to collect at least £20 for Oxfam. How many more miles would she have had to walk in order to do this? *(3)*

▲ *MEG, Level 2/3*

Close study reveals interesting differences among the groups. In a NEA question (not reproduced) Mr Jones buys insurance for a holiday with his wife and three children (surely reinforcing the sexist notion that men have control of the family finances). SEG, on the other hand, has a family book a holiday to Austria. Looking through the MEG specimen questions it was difficult to find one which was not entirely couched in words. Even a gas bill problem was words only, instead of being illustrated by a picture of the actual bill, and requiring candidates to find some information from this. LEAG gets a black mark for continuing to set a multiple-choice paper, but they score well in their specimen questions.

These sample questions have also served to illustrate the fact that it is not in the area of curriculum *content* that we find major differences between existing mathematics examinations and the GCSE.

COURSEWORK AND ORAL WORK

The inclusion of coursework and oral work in all schemes of assessment from 1991 (and optionally from 1988) is a fundamental change, however, and one to which the various groups have responded very differently. The arrangements for oral assessment, especially, differ very markedly from group to group. Three groups only include an aural test, and one of them, LEAG, terms it a 'mental test'. As is explained in Chapter 1, this is a test in which pupils listen to questions read by the teacher and respond in writing. This is very different from an oral assessment in which pupils are required to talk about mathematics!

The summary table (Figure 1) includes the proportion of the total marks allocated to coursework and oral tests in syllabuses which offer this option. Further information is set out below.

LEAG Five tasks are required in the three categories: pure investigations; problems; and practical work. At least one, and not more than two, must be done from each category. Candidates may do tasks set by LEAG or submit their own. Each task is allocated 5 per cent and a further 5 per cent allocated to a mental task, one being set for each level. Oral assessment of coursework is not specifically included.

MEG Five tasks, one from each of the categories, are required: practical geometry; everyday applications; statistics and/or probability; an investigation; a centre-approved topic. Each task is allocated 5 per cent. There is no oral test, but oral assessment can be included in the marking of coursework tasks.

SEG Three units of centre-based assessment are required, one to be an extended piece of work. Total marks are 40 per cent: 16 per cent for the extended piece of work, 8 per cent for each of the other units and 8 per cent for oral assessment of the coursework. A further 10 per cent are allocated to an aural test.

WJEC Two coursework assignments are required: a practical investigation and a problem-solving investigation – set by the board with an option for individual candidates to choose their own topic(s). Total marks are 26 per cent (70 marks out of 270). Practical investigation (30 marks) to include aural test (further 10 marks); oral assessment is allocated 9 marks (out of 30) for each investigation.

NEA Coursework is not specified but guidance is given; it is allocated 25 per cent of the total subdivided (out of 100) as: practical work 30; investigational work 40; assimilation 30 (to include 7 for oral communication).

There are considerable differences among groups in: the weighting given to coursework; the number of tasks required; the categorisation of the tasks; the amount of guidance given and free choice allowed; and the level of oral assessment required/allowed. Three groups have aural (or mental) tests, in one case (WJEC) linked to coursework. (Quite how this would work if candidates did choose their own topics is not clear.) NEA wish to have an arrangement whereby coursework assessment can only be used to improve a candidate's overall grade (that is, if a candidate's coursework mark is worse than his or her examination mark, this would be ignored). The Secondary Examinations Council are not happy with this and, at time of writing, the issue had not been resolved.

Chapter 3 looks closely at the elements which could form part of coursework requirements, and Chapter 5 considers the assessment of these.

3 In the Classroom

> 'Can you do Addition?' the White Queen asked. 'What's one and one and one and one and one and one and one and one and one and one?' 'I don't know,' said Alice. 'I lost count.'
>
> *Lewis Carroll*

CLASSROOM APPROACHES

The single most frequently quoted paragraph in the Cockcroft Report (DES, 1982) states:

> 243 Mathematics teaching at all levels should include opportunities for
> - exposition by the teacher;
> - discussion between teacher and pupils and between pupils themselves;
> - appropriate practical work;
> - consolidation and practice of fundamental skills and routines;
> - problem solving, including the application of mathematics to everyday situations;
> - investigational work.
>
> In setting out this list we are aware that we are not saying anything which has not already been said many times and over many years. The list which we have given has appeared, by implication if not explicitly, in official reports, DES publications, HMI discussion papers and the journals and publications of the professional mathematical associations.

The six items listed in this paragraph were then reiterated in the HMI discussion document *Mathematics 5–16* (DES, 1985). In Chapter 4 they comprise six of the twelve principles which are enumerated and elaborated as the principles which should govern classroom approaches. It is therefore not surprising that all these have been incorporated, in one way or another, into the general and subject-specific criteria for GCSE mathematics. (Another of the principles set out in *Mathematics 5–16* which has been explicitly included in the national criteria is the one which concerns the encouragement of all pupils of both sexes and of different cultural backgrounds. This is discussed in previous chapters.)

As Cockcroft's paragraph 243 points out, this is not a new list. The Cockcroft Committee reminds us that the views expressed there are well documented. There is a considerable consensus amongst people concerned with mathematics education as to the value and importance of all these varied classroom approaches in mathematics teaching. These views arise, at least in part, because it seems that mathematics teaching has not yet succeeded in achieving a numerate and mathematically confident adult population. This is not a new problem, nor is it one which is any worse today than in the past (if anything, there have been improvements over time). The low (and ever deteriorating, it is claimed!) standard of numeracy has given generation after generation cause for concern. So has the apparent inability of students to apply any arithmetical knowledge they might have acquired. For example:

> In arithmetic, I regret to say worse results than ever before have been obtained ... (HM Inspectors' report, Stafford & Derby, 1876)

and:

> Even the enthusiastic apprentice seems to find it difficult to connect the instruction usually given in arithmetic, algebra and geometry with the work of the shop or factory. (Castle, 1899)

Yet another instance:

> It has been said, for instance, that accuracy in the manipulation of figures does not reach the same standard which was reached twenty years ago. (Board of Education Report, 1925)

And more recently, this time from Godfrey (of Godfrey & Siddons):

> Cases of this kind are typical: A boy who is practising decimals in mathematics is found unable to divide by 1000 in the laboratory; he may be studying cylinders in mathematics but breaks down over the sectional area of a cylinder in the workshops... (Godfrey, 1946)

A vision of mathematics education where enjoyment and the confident application of skills and knowledge take the place of fear, panic and the inability to make enough sense of what has been learned to apply it sensibly, is at least as old as these complaints. Nonetheless, paragraph 243 continues:

> Yet we are aware that although there are some classrooms in which the teaching includes, as a matter of course, all the elements which we have listed, there are still many in which the mathematics teaching does not include even a majority of these elements.

The situation today

In most classrooms, exposition by the teacher and the consolidation and practice of fundamental skills and routines dominate to the exclusion of virtually everything else. Discussion, if it happens at all, consists of teacher-led question-and-answer sessions with no opportunity for genuinely open debate or exploration. Conscientious teachers, dealing with large syllabuses, have understandably (although perhaps misguidedly) responded by trying to teach in what they believed was the most efficient manner. More ground could be covered, they felt, in the short time available by using techniques of telling and explaining, and naturally most attention was directed to aspects which were going to be examined, hence the emphasis on practising skills and routines.

The debate on 'understanding in mathematics' exemplified by the article by Richard Skemp, 'Relational Understanding and Instrumental Understanding in Mathematics' (1976), led to much agonising over whether one's pupils were necessarily well served by one's insisting on relational understanding. For many purposes instrumental understanding was quicker, easier, apparently more efficient and often demanded by pupils. How many of us have not had the experience of carefully leading a class towards the 'discovery' of some important mathematical property – the relationship between the diameter and circumference of a circle, for example – only to have them complain, at the end of several lessons when at last the formula had been arrived at, 'Why didn't you tell us that straightaway, Miss?' Pupils, equally, felt under pressure of time. As their experience often led them to believe that none of it would make sense anyway, why not just be given the formula, told how to use it, and then be left to get on with it?

The GCSE will, it is hoped, achieve two very important things for mathematics education. Smaller syllabuses mean that it is now possible to introduce new ideas at a slow enough pace to enable pupils to develop understanding and allow time and space for real discussion and exploration. The importance attached to applications means that key ideas can be approached via a variety of contexts, thus increasing the probability that relational understanding will develop. Including oral work, practical work and investigational work within the formally assessed components, and emphasising real-life applications in syllabus content and therefore in examination questions, gives these elements the status they need if they are to be accorded a proper place in the classroom. Some instances of real-life applications can be seen in the questions which are reproduced in Chapter 2. For example, choosing the best-value coffee jar, with the question being almost entirely pictorial, just as it would be in the shop; buying insurance or carpeting, and having to work out possibilities from the advertisements, etc.

Given all the constraints and pressures, it is not surprising that mathematics education has been slow to change. However, another problem for many teachers, and one which merely the introduction of the GCSE will not immediately change, is the relative lack of experience amongst people working within mathematics education of the different methods of classroom organisation which are appropriate to different needs. How do you plan a discussion lesson? How can you tell whether it has been successful? When is an investigation finished? What *is* practical work? What about discipline? If you've got everyone working in groups, how can you tell whether they're applying themselves to the task or discussing last night's episode of *Dallas?* How can you set homework if everyone is doing different things?

These are very real problems. Many people who wish to teach in different ways have not known where to go for guidance. Traditionally, lesson plans have followed a particular layout, one which is appropriate for a lesson which consists of exposition by the teacher followed by examples for the pupils to practice. A good lesson plan will set out aims and objectives, method, examples to be used and materials required. There is plenty of experience to draw on for lessons like this. When I was a student on teaching practice, an experienced teacher showed me how to improve my lesson plans by sketching out what the blackboard should look like at the end of my exposition, and what the pupils' exercise books should look like at the end of the lesson. Very useful advice. And everyone, parents included, knows what a successful lesson of this sort is like – after all, it's the way we were all taught ourselves! The equivalent experience is not yet readily available for other kinds of lessons, and most of us find it difficult to work with pupils in ways we have not experienced ourselves.

THE INFLUENCE OF CURRICULUM DEVELOPMENT PROJECTS

This lack of experience is, however, only relative and is rapidly being made good. A lot of experience has already been amassed by teachers working within curriculum development projects such as SMILE (Secondary Mathematics Individualised Learning Experience). SMILE is a resource-based learning system and is used by a number of schools in different parts of the country. Pupils in SMILE schools have for some time been able to take a Mode 3 SMILE examination (CSE since 1976 and O level since 1981), and the assessment of these has included both coursework and an investigation paper. Researchers at the Shell Centre for Mathematical Education at Nottingham University, in conjunction with the Joint Metriculation Board, have been working for some time on a curriculum development project. The outcome of this is to be a series of resource packs for schools. At the time of writing two of these packs have been published and provide a wealth of classroom material, including assessment materials and guidance on lesson plans appropriate to the 'missing' elements of Cockcroft Paragraph 243. The GAIM project (Graded Assessment in Mathematics) based at King's College, University of London, is trialling classroom materials in several dozen schools – again material which will be useful for GCSE teachers. And there are other projects and many other sources, not least of which are the continuing debates and articles on aspects of classroom practice published in the journals of the professional associations.

The demands of the GCSE will undoubtedly speed up the dissemination of suitable teaching ideas and codes of practice appropriate to these new methods of organisation and new expectations. Many of the resources already available, including magazines and journals where the 'state-of-the-art' is continually being discussed, reviewed and brought up to date, are referred to (and sometimes quoted from) in various parts of this book, particularly in this chapter. They are all brought together, with publication information and addresses where appropriate, at the end of the book.

The changes in classroom practice and organisation which will be needed if the demands of the GCSE are to be met are discussed in detail in this chapter, with particular emphasis on the 'new' elements – oral work, problem solving and investigational work, and practical work. The role of the microcomputer is also considered.

But first, a brief response to a question which is being asked by many teachers and schools – namely, 'Do we have to replace *all* our existing resources? All those textbooks, bought at great expense and still in good condition – can we still use them?'

NEW BOOKS FOR OLD?

'Will I need lots of new books? And what about different books for the different levels?'

As indicated in the previous chapters, there is nothing in the *content* of GCSE syllabuses which does not appear in existing syllabuses. Lower and middle-level candidates will study little more than List 1 or Lists 1 and 2 together of the national criteria for mathematics, and existing textbooks more than adequately cover this content. It is suggested, in Chapter 2, that schools would be well advised to choose a GCSE syllabus where the additional topics for the top-level candidates are in the main those which have been part of their O level syllabus. If you have done this, then your existing textbooks will cover most, if not all, of the extension material as well.

There is absolutely no need to replace these textbooks. You may on occasion want to approach a topic slightly differently, or offer some practical examples which the textbook does not provide. With judicious selection, and some supplementation, you will be able to use your class sets of texts for a great deal of your work.

As we have seen, the new syllabuses are built from the 'bottom up', i.e. by adding on to a core. This means that all pupils, whatever level they eventually take, will have covered some topics. Extensions occur, as is described in Chapter 2, in various ways These are: additions to an existing topic, more difficult applications of the same basic material and further topics. As time goes on it will become clear which pupils can cope with the more difficult applications or can go further in a particular topic. These pupils can then be offered experience of additional topics as well. But because of the common core, these decisions can be delayed until relatively late in the pupils' secondary school career. In the meantime, activities which extend the work that all are tackling, to provide appropriately 'stretching' experience for some, will be adequate. If work is tackled in an investigational way on occasion, the pupils will themselves respond by going further when they feel capable of this, or are particularly interested and involved.

An example from a 'traditional' topic may serve to clarify this.

Suppose you have a fourth-year class of 'average' pupils. These are pupils who in the past you would have entered for CSE, and would have expected to achieve perhaps a few high grades, but mostly grades 2 to 4. Your school has chosen the London and East Anglian Group's Syllabus A which does not have a school-based assessment, as you wish to build up your experience of coursework before entering pupils for this. It is too early in the course to decide whether these pupils will be entered for Level X (the lowest level) or Level Y (the middle level). There may even be one or two pupils who will improve sufficiently over the next year to be entered for Level Z. You have therefore decided to follow the Level Y syllabus, which of course includes the Level X syllabus. More difficult topics will be left until later, by which time it should be clearer which pupils ought to cover them. You are about to do some graphical work.

A 'traditional' topic – new style

The Level Y (middle-level) syllabus for the topic labelled 'Graphs' says:

> Cartesian coordinates.
> Interpretation and use of graphs in practical situations, including travel and conversion graphs.
> Drawing graphs from given data and to given scales.
> **Constructing tables of values for given functions which include expressions of the form: $ax + b$, ax^2 and a/x ($x \neq 0$), where a and b are integers.**
> **Drawing and interpretation of related graphs.**
> **The idea of gradient.**
> **Solution of two simultaneous linear equations by graphical methods.**

(The items in bold type are the additional items, over and above those included in the Level X syllabus.)

There is nothing here which you will not find in existing textbooks. The only item which is additional to the content in Lists 1 and 2 is 'Solution of two simultaneous linear equations by graphical methods', and this too, of course, is adequately covered in most textbooks. However, the *National Criteria's* aims and assessment objectives (and therefore those of the examining bodies) suggest an emphasis which is perhaps different from that in your old textbooks. For example:

> 2.4 apply mathematics in everyday situations ...;
> 2.8 use mathematics as a means of communication ...;
> 3.1 recall, apply and interpret mathematical knowledge in the context of everyday situations;
> 3.3 organise, interpret and present information accurately in written, tabular, graphical and diagrammatic forms.

In the light of these statements you may find that you need to provide more examples than your textbooks contain of graphs in use in everyday and practical situations, provide opportunities for pupils to develop their own judgement about appropriate graphical presentation of information, and encourage rather more discussion of the interpretation and use of various forms of graphical representation than you might have found necessary in the past.

Studying the specimen examination papers will also help you get the feel of the sort of questions which might be asked.

Remember that Paper 2 will be taken by both Level X and Level Y pupils. Using the standard here as your guide will give you a good starting point. The questions will, of course, only draw on Level X content.

Specimen Paper 2 provided by LEAG has several graphical interpretation questions, two of which are reproduced as Figure 1 and Figure 2.

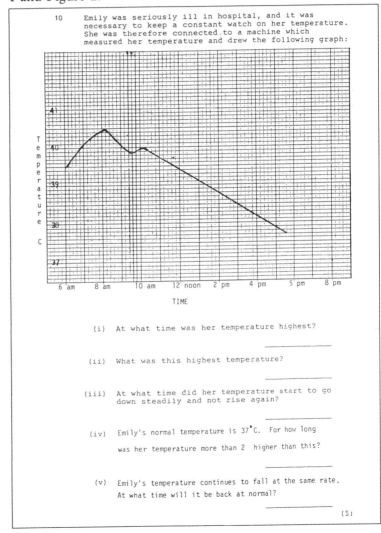

▲ *Fig. 1*

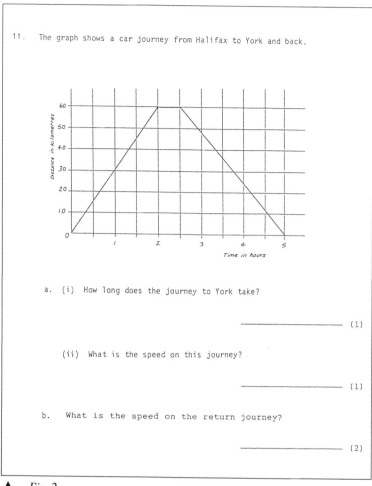

▲ *Fig. 2*

There is also a temperature conversion graph question which requires the ability to plot a graph from given data and then use the graph to obtain further information.

Any of these questions could have appeared in a CSE paper, but it is unlikely that three such questions, and none requiring purely technical knowledge, would have appeared in the same paper. The change is simply one of emphasis, not content.

A quite difficult graphical interpretation question, which one can see as a natural development from the lower-level questions above, was provided by NEA on Paper 4 (i.e. Level 3) – see Figure 3.

Ideas for extension work

Once you start thinking along similar lines, many examples will occur to you and, through discussion with your pupils, further ideas will be generated. A few starters are: A conversion chart for centimetres and inches so that pupils who know their height in feet and inches can easily work out their height in metric measurements; a temperature conversion chart on a scale

2. The diagrams show vertical cross-sections of two cylindrical containers, A and B, and of two containers, C and D, each of which is part of a cone.

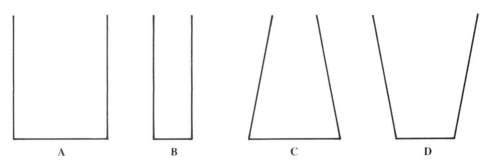

Each container is filled from a tap from which water is flowing at a constant rate. The graphs below show the depth of water measured against time in each of three of the containers.

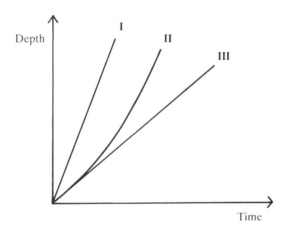

Identify the container to which each graph refers.

Graph I ....

Graph II ....

Graph III ....

▲ *Fig. 3 NEA, Paper 4*

suitable for cooking, so that recipes given in degrees Fahrenheit can be converted to Celsius (or vice versa); lots of examples of travel graphs, including ones involving walking or cycling (with rests, or difficult sections of terrain where one has to slow down); graphs which show baths or swimming pools being filled with water, and so on.

Graphs used in newspapers, magazines and other printed material – advertising brochures and publicity leaflets for example – can be particularly useful. Some of these publicly available graphs offer excellent material for discussion and interpretation. Sometimes, unfortunately, they offer material for discussion on how graphs can be deliberately designed to be misleading and the sort of principles which need to be employed if one does not seek to deceive!

Graphs which relate to the subject matter of other curricular areas are well worth including – all the sciences and technical subjects, of course, but also other subjects such as home economics and the humanities can provide you with examples. This would also be in the spirit of paragraph 19(k) of the *General Criteria* (see Chapter 1, Figure 10).

Looking through your textbooks with a careful eye will undoubtedly reveal more ideas, sometimes in disguised forms.

So the simple answer to the question posed above is, no, do not throw away your old books, and do not go out and buy comprehensive new GCSE mathematics texts – unless of course your books were due for replacement anyway. Most of the content of GCSE textbooks will necessarily be exactly the same as many existing texts. However, you will need to supplement by provision of additional resources of one sort or another.

A RESOURCE PACK: The Language of Functions and Graphs

A most useful additional resource for this particular area of the curriculum is one of the two packs produced by the Joint Matriculation Board and the Shell Centre for Mathematical Education, which are referred to above. Although *The Language of Functions and Graphs* (see Bibliography for further details) was developed prior to GCSE, it is particularly appropriate to support the new elements in the GCSE requirements.

The pack consists of a teacher's book, worksheet masters, microcomputer materials and a video of work in the classroom. The materials have been well tried and tested in classroom situations by teachers and researchers. The book includes a section containing seven specimen examination questions, each of which is accompanied by a full marking scheme, and is illustrated with sample scripts.

In addition to the worksheet masters, the classroom materials include teaching notes. These consist of detailed suggestions for presentation, and on managing and promoting discussions between pupils as well as transcripts of actual classroom interactions. The authors are careful to point out that:

> ... all the teaching suggestions are offered in the recognition that every teacher will work in their classroom in their own individual way. The trials of the material established that teachers found it helpful to have explicit detailed suggestions which they could choose from and modify.

Also provided are additional problems, solutions to some problems and further activities suitable for supplementary work or homework. These classroom materials are an intensely practical, imaginative, well documented and varied resource.

As well as the support materials already mentioned (the video and the microcomputer programs and notes) the book contains further support materials in the form of discussion of some of the important aspects of the new ways of working which can be tackled by means of the materials offered. Headings include 'Tackling a problem in a group' and 'Ways of working in the classroom'.

The workcard about the bus-stop queue (Figure 4) is taken from Unit A. This unit

> ... contains a series of lesson suggestions which focus on the qualitative meaning of graphs set in realistic contexts, rather than on abstract technical skills associated with choosing scales, plotting points and drawing curves. (p. 60)

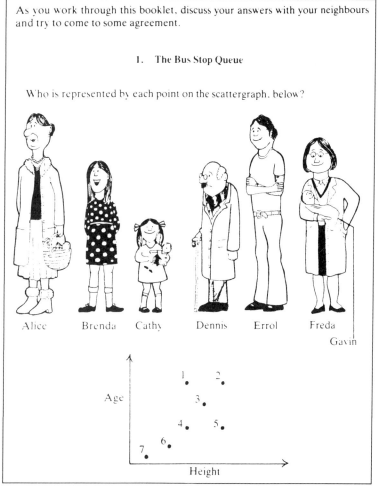

▲ *Fig. 4*

As the authors point out, most courses already thoroughly cover these technical skills and therefore, as is suggested above, existing textbooks will undoubtedly have ample material for developing them.

The aim of the set of workcards is stated as:

> ... to offer pupils an opportunity to discuss and reason about the qualitative meaning of points in the cartesian plane (p. 65)

and a very useful aspect of the teaching notes is the discussion of difficulties which experience has shown may emerge, such as a pupil commenting that, 'There are no numbers on the axes!'

---

ii) Small group discussion

After the problem has been introduced the children are usually asked to work in pairs or small groups. At the beginning of a new task, it often takes some time to absorb all the information and ideas. The group discussions at the beginning of the task may therefore be fragmentary, using keywords, half sentences, questions and so on. We refer to this as the exploratory discussion stage. Although it often appears somewhat disjointed and poorly articulated if the group is left to work undisturbed, it is here that organising and reformulation can emerge.

If you are able to tape record some small group discussions you may like to analyse them in the following way:

1. Divide the discussion into a number of distinct episodes or subtopics, as self-contained as possible.
   Identify the initiator of the episode, and discover whether the initiator is a group member or leader (teacher).

2. Can you find examples of participants:

   a) putting forward a tentative or hypothetical idea, and asking for comment?
   b) supporting their own assertions with evidence?
   c) contributing evidence in favour of someone else's assertion?
   d) pointing out flaws in the arguments or questioning 'facts' put forward by others?

   Are all members of the group:
   e) participating?
   f) supporting the discussion?

3. What kinds of intellectual process were being used? Count the following, putting doubtful cases into more than one category if necessary:

   a) contributions principally at the level of *specific information* (data);
   b) contributions that focus on *ideas* or *concepts* (classes of events, objects or processes);
   c) the number of *abstractions* or *principles* involving more than one concept.

▲ *Fig. 5*

Using the microcomputer (1)

Work with a microcomputer is, rightly, integral to the materials offered. It is of use in so far as it supports the key concepts and skills being taught, or extends them in some way. Mathematics classrooms should have regular and easy access to a microcomputer in order to use it whenever an aspect of the work can be more easily or more satisfactorily covered with it than without it, or when it adds an extra dimension to the activity. 'Using the micro' should not be seen as a separate activity, any more than talking, writing, solving problems, drawing graphs or constructing shapes are separate activities. These are all things which go on as a part of doing and learing mathematics.

One way in which the micro is particularly helpful is in changing the power relations in the classroom from one where 'teacher has all the answers, so our job is to work out what she or he is thinking' to one where everyone in the room is trying to work something out, teacher included. One of the micro programs provided in the pack would help develop the kinds of concept which are needed to solve the bottle-filling question above and could help transfer control of the learning situation from teacher to pupils. It generates bottles of different shapes and sizes, filling with water, and asks how the graph changes or asks the user to work out the shape of the bottle when the graph is provided. This program could be used for class work and discussion or for small-group work.

Classroom discussions

Transcripts of discussions in lessons provide a most interesting dimension of the materials in the pack. Some analyses of small-group discussions are provided, together with guidelines for analysing tape recordings that teachers may be able to make in their own classrooms. Page 221 of the book (reproduced as Figure 5) makes some important points about group discussion and provides a framework for analysis. This framework has been used to analyse the transcript of a discussion about the bus-stop problem which is reproduced as Figure 6.

This leads naturally into the next section of the chapter.

ORAL WORK IN MATHEMATICS

Readers will recall that oral work in mathematics is among the 'new' elements which GCSE will be examining, in one way or another, by no later than 1991. Assessment objective 3.16 states that any scheme of assessment will test the ability of candidates to 'respond orally to questions about mathematics, discuss mathematical ideas and carry out mental calculations.'

Chapter 2 outlines the responses of the various examining groups to this requirement. It was found that these were very varied, ranging from SEG at one extreme to MEG at the other. (These remarks only apply, of course, to the syllabus provided by each group which includes a coursework element.) SEG has devised an assessment scheme for its syllabus in which 50 per

Worksheet A1   1. The Bus Stop Queue (Three boys: P1, P2, P3)

	Transcript		Category and Comment
P1	Right. Obviously the two highest are Alice and Errol.	1 3a	Initiator is P1 not the teacher Specific information contributed
P2 P1 P2	Yeah. Numbers 1 and 2 are both the tallest. Yes. Therefore they're Alice and Errol.	2b	P2 makes an assertion, but it is based on the misconception that 'high points' = 'tall people':
P1 P2	Hold on! No! 1 and 2 are both the two oldest. They're Errol And Alice. Yeah. That's what I said.	2d	P1 points out a flaw in P2's argument, but then makes a slip himself.
P1	Sorry . . . I think it could be Dennis and Alice?	2a	P1 puts forward a tentative (correct) idea.
P2	But Dennis is shorter.	2d	P2 questions P1's conclusion.
P1 P2	How do you know that Freda isn't older then? Don't be silly. Use your common sense.	2d	P1 seems to be trying to point out a flaw in P2's argument by questioning.
P1	Um . . . so Alice'll be the older one. So Alice'll be number 2. OK?	2a	P1 returns to his own approach and asks for comment.
P2	What? She's the oldest and she's the tallest?	2d	P2 implies that there is a flaw in P1's argument.
P1 P2	The other oldest one is short, so that's number 1 isn't it and that's Dennis. Hey up will you two do something? Well it says agree and I'm agreeing!	2b 3b 2e 2f	P1 supports his assertion with evidence. P1 looks at both variables simultaneously. P1 feels that he is doing most of the work! although P3 was silent during the episode, he was supportive and involved.

▲ *Fig. 6*

cent of the total marks are awarded for the coursework element. Two aural tests total 10 per cent and three units of centre-based assessment total 40 per cent. One-fifth of this latter 40 per cent is allocated to oral assessment based on a series of discussions about the candidates' work. Thus aural tests and oral assessment together amount to 18 per cent of the total. MEG on the other hand allocates only 25 per cent to the coursework element, does not include an aural (or mental) test, and only lists 'an oral exchange between the candidate and the teacher' as one possible means of assessing candidates' coursework assignments. There is a full discussion of aural tests and oral assessment in Chapter 5.

There are two distinct elements encapsulated in the brief statement of objective 3.16. The first concerns the ability 'to respond to questions about mathematics' and the second to the ability to 'carry out mental calculations'. Calculations are, however, only one of the many kinds of mathematical activity which can be done 'mentally'. And the ability to respond orally to questions posed by a teacher or examiner is only a part of the whole area of discussion of mathematical ideas. It seems a pity that the national criteria did not specifically mention such things as the ability to visualise in three dimensions, the ability to pose questions, or other aspects of what could usefully be included in oral and mental mathematics. Perhaps these will be incorporated at a later stage.

The importance of oral work is, however, not merely a function of how many marks are specifically allocated to it in an assessment scheme. The importance of language – reading, writing, talking and listening – in the learning of mathematics has gained general recognition in recent years, and can be seen as part of the movement towards language across the curriculum, which the Bullock Report (DES, 1975) exemplifies.

The old aphorism, 'How can I know what I am thinking until I hear what I have said?' may be exaggerated, but nonetheless contains an important element of truth. The relationship between language and thought is a complex one, which can perhaps be illustrated by a Venn diagram (Figure 7). It is possible to have thought without language, as some animal studies (e.g. Köhler, 1925) and studies of patients suffering from aphasia (loss of language owing to brain damage) have shown. It is also possible to have language without thought, as in babbling. Most thought however is conducted through the medium of language, and it is through communication, with others and privately, orally or in writing, that ideas are clarified and made our own.

The trouble with most mathematics classrooms is that a large proportion of the talk is teacher-talk. Certainly, the typical

▼ *Fig. 7 The relationship between language and thinking*

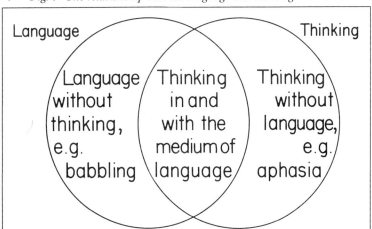

traditional lesson described above of exposition by the teacher followed by practice of examples by the pupils leaves no space for children to explore concepts and ideas through discussion. Even if the teacher is conducting the exposition by means of questions and answers, and thinks of this as a discussion, it is likely that he or she will be doing most of the talking and will decide whether or not a pupil's response is acceptable or valuable. A lot of the time the pupils are likely to be engaged in a game of 'guessing what's in the teacher's mind' rather than applying themselves to the problem and thinking independently. There is a vast difference between the kind of 'discussion' which is teacher-led, and is going to go in the way the teacher directs, and open-ended discussion.

The example (Figure 6) of pupils discussing the bus-stop problem is an instance of open-ended discussion which is not being directed by the teacher. The teacher has set the situation up, and then left the pupils to explore this without her presence. Discussion involving the whole class can also be open-ended, if the teacher's role is that of chairperson. In most classrooms, however, teacher-directed discussion is the norm.

An example of a teacher-led discussion

Many of these points are made by the authors of a book devoted to language in mathematics education (Harvey, 1982), where the assertions are backed up with the results of research into classroom interactions. A number of transcripts of classroom talk are included and are very illuminating. The extract below is taken from the chapter by Daphne Kerslake.

1	T	What did you notice about the fractions (a) (3/4, 9/12, 30/40, 33/44, (b) (5/8, 10/16, 15/24), (c) (8/12, 12/18, 10/15) when you worked them with the calculators?
2	C1	They're all the same.
3	T	What is the same as what?
4	C2	They're just the same.
5	T	What are?
6	C1	In (a) they're all the same. In (b) they're all the same.
7	T	Can you see why they're the same?
8	C2	3 will go into the top ones and 4 will go into the bottom ones.
9	T	Yes ... is there anything else?
10	C	...
11	T	Suppose one of the ones in (a) had been 9/80? ... You could say '3 goes into 9 and 4 goes into 80'?
12	CT	Yes.
13	T	OK. Try it with the calculator.
14	C2	Oh no! ...
15	T	So when you said '3 goes into the top numbers and 4 goes into the bottom ones', you were right, but you need something more ...
16	C	...

17	T	Would 9/16 work?
18	C2,C4	... No.
19	T	But 3 goes into 9 and 4 goes into 16. So 9/12 does work, but 9/16 doesn't ...
20	C1	3 goes into 9 three times, and 4 goes into 12 three times. It's times by 3.
21	T	Yes, good. And with 33/44, 3 goes into 33 eleven times and 4 goes into 44 eleven times. So what about the next ones – in (b)?
22	C1	5/8, 10/16 ... 5 goes into 10 twice and 8 goes into 16 twice.
23	T	Good. And 5 goes into 15 ... C2?
24	C2	3 times, and 8 into 24 three times.
25	T	Fine. What about the third set? 8/12, 12/18, 10/15?
26	C3	8 and 4 is 12, 12 and 6 is 18.
27	T	Um ... C1?
28	C1	4 goes into 8 twice, 6 goes into 12 two times, 4 goes into 12 three times, 6 goes into 18 three times.
29	T	Good. Well done. Could you find another way of writing 8/12 that would give the same number on the calculator?
30	C1	4/6.
31	T	And an easier one still?
32	C2	2/3.
33	T	So 2/3 and 12/18 are the same because 2 goes into 12 six times and 3 goes into 18 six times. (pp, 63–5)

Daphne Kerslake points out the extent to which the teacher is directing this interchange, how few of the fifteen pupils actually participate in the discussion and how little evidence the teacher has at the end as to what most of the children understand. She also comments that this teacher

> prides herself on her ability to draw ideas out of her class and not intervene too much. Before listening to the tape she commented: 'I thought the children contributed well and I think they really began to understand the idea of the equivalence of fractions for the first time.'

It is indeed very difficult to be really objective about ourselves in the classroom, and it is often a sobering and salutary experience to see ourselves on video or hear ourselves on tape working with children.

Analysing oral work

An important piece of research cited in Daphne Kerslake's chapter is that of Flanders (1970), who devised a method of analysing classroom dialogue. Ten categories of talk were identified, seven for teachers and two for pupils, with the tenth category being silence or confusion. The seven teacher categories were subdivided into 'direct influence' (lectures, gives directions and criticises or justifies authority) and 'indirect influence' (accepts feelings, praises or encourages, accepts or uses students' ideas, asks questions). The two categories of pupil-talk were response to the teacher and talk initiated by the student. The result of many classroom observations where talk was coded according to these ten categories led Flanders to suggest a *rule of two-thirds:*

> In the average classroom someone is talking for two-thirds of the time; two-thirds of the talk is teacher-talk, and two-thirds of the teacher-talk is direct influence.

The transcript above certainly demonstrates a teacher taking up two-thirds of the talk time. Allowing for the difficulty of counting numbers of words in a mathematical exchange, the totals are 193 words spoken by the teacher, and 109 words spoken by all the pupils together.

There is also substantial evidence that in mixed-sex classrooms boys' talk takes up a majority of the pupils' talk time. It is this evidence, amongst other things, which led many people who are concerned about gender equality to suggest that girls would benefit from being in single sex-schools, or at least single-sex teaching groups for mathematics and science.

Most classrooms, then, do not offer pupils adequate opportunity to make ideas their own through talk. And mixed-sex classrooms generally offer even less opportunity to girls than to boys.

It is not easy to alter classrooms so as to encourage talk by pupils, not least because it is so ingrained into teachers that they are not doing their jobs properly unless they are explaining things to pupils. One of the hardest thing for a teacher is deliberately to hand over some of the control for pupils' learning to the children themselves. This must be done, however, if improvements are to be achieved in children's learning of mathematics.

Working in small groups

As a mathematics education community we are gradually learning more about the sorts of classroom organisation which facilitate learning through talk. One of the most effective methods of organisation is groups of two to four learners, optimally probably two or three people in a group. Larger groups sometimes work, but too often they result in one or more individuals being on the fringe of the learning situation, contributing and learning little. Even in the bus-stop problem discussion (Figure 6), which involved only three pupils, one of them was silent during the episode. The comment says that he was 'supportive and involved,' although we are not told the evidence for this. The bigger the group, the harder it would be (a) for everyone to participate actively and (b) to judge whether silent group members were involved in what was going on.

You might like to try out a lesson in which you organise your class into small groups, which may or may not be self-selected. Pose a problem for discussion, preferably one to which there is not just one correct answer. Make it clear that this is the case and that you are setting this up to give pupils an opportunity to discuss the problem amongst themselves rather than to 'find an answer'. You could use some of the worksheet ideas discussed under the graphs section above, but if you are not currently planning to do graphical work you might prefer a self-contained problem which is not too dependent on specific background knowledge. A good example to use might be the 'Design a kitchen' problem from the SMILE bank of materials. Another is the job-sheet problem, Winston's garage, which is a GAIM resource. These are reproduced in Chapter 4.

Both problems would be suitable material for a class (and teacher) new to this way of working. The problems are practical and relate to familiar situations. In both cases it is obvious that there may be more than one acceptable solution. Equally, it is obvious that talking things through with other people and listening to their ideas could be helpful in finding alternative solutions. Which solution is 'best' in both cases depends on assumptions or on aspects of preference and aesthetics. A pupils's ideas are therefore just as likely to be useful as a teacher's ideas. Nonetheless, there are clearly defined constraints which mean that some ideas will not work. Unworkable schemes can be pointed out to pupils by each other without teacher intervention being inevitable. The problems are self-contained and they are designed to be completed in a fairly short time.

You could try out one or other of these problems, in an experimental fashion, with one of your younger classes. Choose a class with whom you have a comfortable relationship and no discipline problems, and take them into your confidence. Tell them that one of the things they need to learn is how to discuss mathematical problems, and that you want to try out different ways of working to make it easier for this to happen. Discussing mathematical problems involves listening to other people's ideas as well as explaining their own ideas to others. In order to give as many people as possible the opportunity to do this, you want them to work in small groups on a problem to which there are several possible solutions. (I recommend groups of three to start with – you can amend this in the light of experience.) Emphasise that what matters is talking things through with others in their group rather than coming up with 'the answer'.

Now comes the hardest part – let them get on with it! Look out of the window if necessary, or do some pencil sharpening. After a few minutes, stroll around the room and join a group. Try to be a listener rather than a teller. Inevitably, the children will ask 'Is this right?' Ask them what they think – remind them there is no one right answer. Try to turn yourself into a resource, to be used to supply necessary information, but not a decision-maker. Remember that, if you are talking, your pupils are, at best, listening. If they are talking, not only is the person doing the talking also thinking, but there is a greater likelihood that the others are listening properly and thinking about what is said. After all, teacher is sure to be correct, so we do not need to examine what she or he says, we just have to accept it. Our fellow learners, on the other hand, might well be wrong, so it is worth listening carefully enough to decide for ourselves whether or not they are right.

If you hear someone going badly wrong, try asking a question such as, 'Can you explain that?' or 'Why do you think that?' rather than directly correcting the misapprehension. Quite often, children will see for themselves that they are going wrong and will correct it – if the work is within their capabilities. Also, by listening to their responses you will learn a great deal about what the children know and understand and where their difficulties lie. You can then plan activities for them which provide missing experiences and knowledge or will help them to make necessary connections. As Daphne Kerslake says in the book cited above (Harvey, 1982), 'listening to children is really the only way to find out what is going on in their minds' (p. 71).

Using the microcomputer (2)

The microcomputer can be a very useful resource for getting good, open-ended discussion going in the mathematics classroom. It is now the experience of many teachers that children working in pairs or small groups around the microcomputer talk a great deal, and most of the talk is directed at the problems they are engaged on.

The power of the microcomputer to facilitate discussion among pupils lies partly in the fact that control of the learning situation is to some extent transferred from teacher to pupils. Feedback comes from the micro, and this decides whether one is being successful – at solving a problem, creating a design, getting the snooker ball into the pocket, or whatever. The teacher can be used as resource, a more experienced and knowledgeable person with whom one can discuss ideas or who may have helpful suggestions to make, but the teacher does not, generally, know the answers – if indeed there are any answers! Children exploring, discussing, explaining (to teachers or other pupils) – all with minimum teacher intervention – are common aspects of work with microcomputers.

There are now some excellent programs available for the mathematics classroom, many designed to be tackled by a small group of pupils working together. A number of sources are listed in the Resources section at the end of the book. Attention is also drawn to *Micromath*, a journal for mathematics teachers wishing to use the microcomputer imaginatively in their classrooms.

Problem-solving through programming, whether in LOGO or BASIC, (unpopular though BASIC now is in some circles!) also offers many opportunities for discussion of mathematical ideas. LOGO, an interactive programming language with excellent graphics facilities (usually called 'turtle graphics'), has by now been introduced into many classrooms, for learners of all ages from 4 to 84! Teachers who have seen the quality of children's mathematical learning when they program in LOGO generally need no further convincing. A great deal has been written about LOGO since Seymour Papert's seminal book *Mindstorms* (1980) was first published. Through the medium of LOGO fruitful oral, investigational and aesthetically creative work as well as highly logical and structured thinking can evolve, with minimum intervention from teachers. The mini-transcript in cartoon form (Figure 8) is an amusing example of a LOGO interaction.

Even when the micro is being used as a whole-class resource, perhaps to display graphs for discussion, the fact that the problems are set by the program and not the teacher means that teachers can take 'a back seat' (literally and figuratively) and work with pupils to solve a common problem.

Currently (1985–6) the program 'L – A Mathemagical Adventure', an adventure game in which Runia tries to escape from Drogo robots by solving mathematical problems, is creating a great deal of delight and excitement in mathematics

▼ *Fig. 8, reproduced from Mathematics Teaching No.114*

classrooms around the country. It has also proved the stimulus for much discussion, practical work and problem-solving and investigational work (of which more below).

'But what about the noise?'

But, many teachers must be thinking, my classroom will be so noisy if everyone is talking – no one will be able to concentrate! I do not wish to minimise this problem, which is also commented on by Tony Purcell in Chapter 4, but I do believe that it is possible to keep the noise down to an acceptable level by persuading pupils to talk in fairly low voices so that the overall effect is a buzz of conversation. This is not easy and takes practice (for teachers and pupils). If the classroom is one where children are expected to take responsibility for their own learning, it will also have a good chance to become one where talk which disturbs others is seen by pupils (and not just the teacher) as antisocial and selfish. Classroom discussion is so vital that techniques must be used which make this possible despite any practical difficulties which might arise.

PROBLEM-SOLVING AND INVESTIGATING

investigate *v.t.* Examine, inquire into, study carefully.
problem (mathematical) *n* inquiry starting from given conditions to investigate a fact, result, or law.

Concise Oxford Dictionary

If you are uncertain as to the difference between 'doing investigations' in mathematics and 'problem-solving' in mathematics, you are not alone! There is a great deal of overlap (not to say confusion) in the use of these terms, and certainly no consistent use has yet emerged in the mathematics education world. Even the Cockcroft Report is unclear, sometimes linking problem-solving with applications (e.g. para. 249, 'The ability to solve problems is at the heart of mathematics' ... etc.) but also talking about 'puzzles and problems' (e.g. para. 226) which gives quite a different flavour to the idea of a problem.

This inconsistency in the use of the terms is strongly reflected in the way the different examining groups state their coursework requirements. (Chapter 2 includes a summary of these.) LEAG divides coursework tasks into three categories which they call pure investigations, problems, and practical work. WJEC requires that candidates tackle two tasks, a practical investigation and a problem-solving investigation. MEG's categories are: practical geometry; an everyday application of mathematics; statistics and/or probability; an investigation; and a centre-approved topic (which can be almost anything). NEA says 'No attempt has been made to distinguish the types of activity which may be included within the general heading "coursework"', and offers guidelines for the award of marks in which the main headings are: practical work and investigational work. MEG also mainly distinguishes between practical work and investigational work.

The dividing line, if indeed one exists, between problem-solving and investigating in mathematics is a very fine one. One commentator, David Wells, in an article published in *Mathematics in School* (Wells, 1985), claims that 'any distinction between problems and investigations is misleading and not followed by professional mathematicians'. Similarly, what constitutes practical work is not clearcut. This is considered later in this chapter.

Here I shall focus on that element of GCSE mathematics which appears, in one guise or another, in each of the examining groups' coursework requirements. Whether it is called 'a problem solving investigation' (WJEC), 'a pure investigation' (LEAG) or simply 'investigational work' or 'an investigation' is not significant, as very similar sorts of activity are intended. So what is meant by these phrases?

Susan Pirie, in the introduction to the pack *Investigations in Your Classroom* (Pirie, 1986) suggests that a valid response to the question 'What is an investigation?' is to start by saying what it is not. She continues:

> It is not a task with a prescribed route to a single solution. It is not an exercise with the overt intention of repetitiously practising a mathematical skill albeit disguised as a word problem. An investigation presents an open situation. For the pupil there are no known outcomes. There may be no known outcomes at all. Pupils are not expected to produce 'the right answer' but are required to explore possibilities, make conjectures and convince themselves and others of what they find. The emphasis is on exploring a piece of mathematics in all directions. The journey, not the destination, is the goal.

A distinction which I find useful and which is quite often made, implicitly if not explicitly, is that an activity with the characteristics that Susan Pirie describes should be called an investigation while a question to which there is a specific answer – although perhaps many routes to the answer – should be called a problem. However, it is necessary to be aware, as David Wells points out in the article cited above, that '"openness" is a function of the problem in relation to the solver, not of the problem alone' (Wells, 1985). In other words the same 'problem' may or may not be, for students at different levels of sophistication, 'an open situation', and hence for them a starting point for investigation.

Active problem-solving

In the section on oral work I state that one of the principal advantages, for pupils, of discussing mathematical problems, especially with one another, is that they are much more likely to be directly engaged in the problem as opposed to when they are just listening to the teacher. The same point can be made about the (relatively) passive exercise of reading about problem-solving and investigating (or reading other people's solution to a problem), in contrast to trying to solve a problem or investigate a

> **Lock–unlock**
> The prison was a long narrow strip of 100 cells. The first warder walked along the corridor unlocking all the cell doors. The second warder went along and locked every second door. The third warder went along the prison corridor and changed the state of every third door, i.e. if the door was locked then he unlocked it, if the door was unlocked then he locked it. The fourth warder next went round the prison and changed the state of every fourth door. The fifth warder changed the state of every fifth door and so on. When the hundredth warder had changed the state of the hundredth door, all the warders went home for tea. *How many prisoners escaped?*

▲ *Fig. 9*

situation for oneself. So readers are strongly recommended to have a go at 'Lock–unlock,' a polyomino or polyominoid investigation, ideally as a group activity with one or more colleagues, before reading beyond this section.

'Lock–unlock' is taken from the problem page of *Investigator* No.3. (Figure 9). *Investigator* is a newspaper produced once a term by teachers through the SMILE centre (see the details in the Resources list). It is a very useful source of problems, investigation starting points and accounts of classroom activities. Several of the teachers who have contributed to this book are also contributors to *Investigator*. 'Lock–unlock' appears under the label 'problem' (not investigation). It does ask for a particular solution and is not obviously open-ended, and so this labelling fits the distinction made above.

Polyominoids and polyominoes

▼ *Fig. 10 A trominoid (top) and below, two of the many tetrominoids*

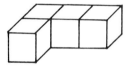

▼ *Fig. 11 There are two distinctly different trominoes*

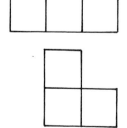

Polyominoids (see Figure 10) are the three-dimensional equivalent of polyominoes, which may be familiar to some readers. The writer first came across polyominoes in Martin Gardner's book *Mathematical Puzzles and Diversions* (1959), where an entire chapter is devoted to them. For readers who are new to them, a brief description follows.

A polyomino is a set of squares which are joined along their edges. A monomino consists of just one square, and dominoes are polyominoes made up of two squares. Quite clearly there is only one distinctively different monomino and one distinctively different domino (assuming that rotations of a domino do not count as distinctively different). There are two trominoes (see Figure 11). By the time one gets to exploring tetrominoes (four squares), pentominoes (five squares) and hexominoes (six squares) the numbers increase rapidly, and other kinds of investigation (in addition to 'How many can you find?') open up. For example, with pentominoes it is possible to investigate which of them will make an open box when cut out and folded up, whether the whole set of different pentominoes will 'jigsaw' into a rectangle (or into several different rectangles), whether any or all will tessellate, and so on. Some hexominoes form nets of cubes – which? and why? – lots of jigsaw and tessellation investigations are possible and so on. If you want some more polyomino ideas, read the Gardner book or *Sources of Mathematical Discovery* (Mottershead, 1978) – or just explore!

If you have never explored polyominoes, you may prefer to do some work on this; otherwise try polyominoids. Lorraine Mottershead also describes polyominoids as follows: 'Allied to polyominoes, polyhexes and polyiamonds (all with only two dimensions, length and breadth) there are three-dimensional shapes called polyominoids. Instead of using unit squares, as in polyominoes, cubes can be joined together face to face. Centicubes are ideal for building up a set of polyominoids.' However, it is strongly recommended that you don't read on until you have explored one or the other and have solved 'Lock–unlock'.

Once you and your colleagues have solved 'Lock–unlock' and 'finished' your investigation of either polyominoes or polyominoids, it would be useful to spend some time discussing the activity. Some questions you might think about are outlined below, and a transcript of two teachers working on 'Lock–unlock' is included.

Reflecting on problem-solving

Some questions to reflect on are: What strategies were used?; Did you make use of (or invent) some form of notation?' Did you use diagrams?, symbols?, representation?; Did you tabulate results or order them in any other way?; Did you use any aids – e.g. squared paper, scissors, ruler, Sellotape, Centicubes?; Did you spot patterns, or notice any short-cuts?; Did you try your thoughts out on one of your colleagues?; Was that a help or a hindrance?; What about any ideas they shared with you – were they a help or a hindrance?

Affective responses

Equally, it is important to talk about and reflect upon how you felt while you were working on the problem. What words best describe your feelings – were you confident? bored? fearful? anxious? resentful? excited? panicky? interested? enthusiastic? Did you find the activity pleasurable? Or was it a chore? Do you think any of your feelings or responses were engendered or exaggerated by your being in a group situation – e.g. were you more confident of success because of having others to share ideas with or, conversely, afraid of being shown up as inadequate relative to others? Or, if you worked on your own, how do you think this might have affected your emotional responses to the problem?

Affective aspects of learning, in other words the effect on learning of the way people feel in a learning situation, can be critically important in determining success or failure. This is particularly relevant to mathematics, as there is plenty of evidence, anecdotal and research, that of all the subjects generally studied at school, mathematics is the one which is most likely to arouse strong, usually negative, feelings in learners. (See, for example, Buxton, 1981; Hoyles, 1982.) Many people either love mathematics or hate it, and sadly, more people hate it or are terrified of it than love and enjoy it. Negative feelings abound, and girls, more than boys, tend to suffer from them. Girls often lack confidence, panic about mathematics and are very dependent on teacher approval. Many researchers, including the writer, believe that this is one of the factors which result in girls' relatively poorer attainment in mathematics. The recently published collection *Girls Into Mathematics Can Go* edited by Leone Burton (1986) contains several papers where this is discussed (e.g. chapters by Lynn Joffe and Derek Foxman, Hilary Shuard).

There is a stated commitment in the GCSE documentation to

the elimination of gender bias in examinations, and some of the specimen questions suggest that examining groups are, quite often, trying to amend the style of question they set so as to make it clear that girls and women are included in the community of people who use mathematics. This is vitally important, but it is equally important to be aware of the possibly somewhat different needs of girls and boys in the classroom. There is beginning to be evidence that less competitive, more collaborative styles of working are beneficial for many pupils, but especially for girls. Our society tends to train boys to be competitive and so they, on the whole, find competition less threatening than girls do.

These observations tie in with the research on classroom talk discussed above – boys are generally more willing to participate in a whole-class situation, while girls often find this intimidating, especially in a mixed-sex classroom. Quite often teachers who are sensitive to pupils' feelings hesitate to ask girls questions when conducting whole-class discussions, as they recognise the girls' discomfort and embarrassment. Unfortunately, although the motivation for avoiding asking girls to respond in a large mixed group is admirable, the outcome is fewer opportunities for girls to try out ideas. Working in a small group gives girls the opportunity to participate more freely and takes away the fear of being 'shown up' in front of a large audience.

However, it is not only competitive situations which arouse negative feelings. Being faced with tasks which seem too difficult, or where one does not know what to do or what is expected, can also arouse negative feelings. It can be very difficult for teachers to get in touch with these feelings in themselves because most of the time the curriculum material being taught is very familiar. It is only occasionally, perhaps while working through an A level examination paper, that most teachers might meet a problem to which the solution – or perhaps the method of attack – is not immediately obvious. However, in open-ended investigating situations we do all find ourselves much closer in spirit to the unsophisticated learner, and it is therefore much easier to relate to the feelings our pupils are likely to be experiencing.

Problem-solving strategies

The transcript below is of two teachers working together on 'Lock–unlock'.

1	S	I suppose we need some way of recording what's happening – should we write down the numbers from 1 to 100 to stand for the cells?
2	E	Yes, let's do that – (pauses, reads question again) – we have to find out how many prisoners escape – I suppose they only escape if their cell is open when the wardens have gone?
3	S	(Busily writing numbers in a rough array, leaving lots of space around them.) That's got to be what it

		means because otherwise they'd all escape straight-away!
4	E	Yes of course ... and that means it doesn't matter how many times a cell is locked and unlocked during the day. All that counts is the final state ...
5	S	I've written the numbers out – let's work out what happens to some of the cells ... it must be to do with factors ...
6	E	... Yes, it must be ... cell 1 will be unlocked, because the warden won't ever go back, and cell 2 will be locked.

(S. writes 'unlocked' above cell 1 and 'locked' above cell 2.)

7	E	Shall we have some shorthand – how about U for unlocked and L for locked?
8	S	Yes, O.K.
9	E	Let's look at the other prime number cells – they'll all be visited just once more, won't they? So they'll end up locked. Is that right?
10	S	Yes, they must be, because they'll all be visited exactly twice. I'll write the factors underneath – I think each cell will be visited as many times as it's got factors.
11	E	That seems right, let's check it out – yes, 4 is visited three times (1, 2 and 4) and 6 is visited four times (1, 2, 3 and 6), 4 will end up unlocked because there are an odd number of visits and 6 will end up locked because it has an even number of visits ...
12	S	So it's connected with whether a number has an odd or an even number of factors ... let's do some more ...

(They work some more out, and find that 8, 10, 12, 14 and 15 are locked while 9 and 16 are unlocked at the end of the day.)

13	E	Writing it all out takes ages, there must be a quicker way – let's stop and look at what we've got ...
14	Both	... Could it be to do with square numbers? All the unlocked cells so far are square numbers, 1, 4, 9 and 16.
15	S	Yes, 25 works as well – it has three factors.
16	E	And so does 49 – I haven't worked out 36 – lots of factors there!
17	S	What about multiples of square numbers? No – 8 is a multiple of 4 and that's no good.
18	E	It must be that only square numbers have an odd number of factors ... but why?
19	S	Yes, that's it – factors of a number are always in pairs, except for the square root of a square number, which is on its own – so that gives you an odd number of factors.
20	E	We've done it!

In the first few interchanges, S. and E. are agreeing on a representation of the problem, and negotiating and clarifying the meaning of the problem. They then try specific examples, and on the way invent a symbolic representation. They quickly see that factors of numbers are involved and that whether a number has an odd or even number of factors is important, but do not see the implication of this yet. The square numbers are spotted initially as 'leaping out of the page', and it takes several more interchanges and some thought before they both understand why square numbers have odd numbers of factors.

We can see at work here many of the elements of problem-solving which are so often written about and discussed – representation and notation; negotiating and agreeing the meaning of a problem; trying out specific examples; criticising strategies because of tedium and trying to find short cuts; looking for patterns; trying further examples to test a conjecture or hypothesis; looking for the underlying reason for a pattern which has been spotted; generalising the solution. Note that E. and S. did not feel they had solved the problem until they had understood why square numbers have an odd number of factors – just spotting the square number pattern and linking this to odd numbers of factors was not enough. They felt the need to generalise once they were reasonably confident of their hypothesis. Less sophisticated problem-solvers might well feel they had 'finished' the problem if they had spotted the square numbers coming up, and checked out that all the square number cells were unlocked and none of the others were.

There are many excellent sources of information and advice on developing problem-solving strategies for ourselves or for our pupils. The classic in this field is probably George Polya's *How to Solve It* (Polya, 1957), first published in 1945 and still relevant today. In his preface to the first printing Polya says:

> Yes, mathematics has two faces; ... Mathematics presented in the Euclidean way appears as a systematic, deductive science; but mathematics in the making appears as an experimental, inductive science. Both aspects are as old as the science of mathematics itself.

It is this second aspect, 'mathematics in the process of being invented', which has tended to be absent from the mathematical experience of most learners and which teachers now, in response to the GCSE changes, will need to share with pupils as part of the day-to-day work of the classroom. 'Lock–unlock' was solved in an experimental and inductive way initially, and only systematised as a very last stage. The problem could have been presented in a formal way, 'find which numbers between 1 and 100 have an odd number of factors and explain your result' and the learners taken through the stages in a systematic, deductive way. But the learning experience would have been much poorer. Not only did E. and S. practise and improve their problem-solving skills; they also ended up with an understanding of factors which is much deeper, much more their own, than if this had been presented as a formal result.

Polya describes four stages in problem-solving:

- Understanding the problem
- Devising a plan
- Carrying out the plan, and
- Looking back.

We can pick out all these stages in the episode above. The middle two stages are returned to several times as E. and S. revise their plan in the light of their growing comprehension of the problem. This is by now a very well known schema for analysing problem-solving which has had a great influence on later work on problem-solving. Another, not dissimilar, schema is provided in two more recent books. John Mason in *Thinking Mathematically* (1982) introduces the idea of three phases of work in problem-solving. These are: entry, attack and review. Mason analyses and illustrates these phases through numerous examples of problems and their solutions. This is an excellent book, full of practical advice, such as on how to make use of being stuck, in order to make progress. Mason emphasises the importance of questioning, challenging and reflecting, and in the final section has this to say about mathematical thinking:

> ... mathematical thinking has a very special contribution to make awareness in that it offers a way of structuring, a direction of approach, a reflective power as well as creative and aesthetic potential. Whether the focus of questioning is practical and related to the material world, or more abstract dealing with numbers, patterns, and structures ... resolving brings a sense of pleasure and confidence.

In contrast to *Thinking Mathematically*, which is directed at readers who wish to improve their own problem-solving skills, Leone Burton in *Thinking Things Through* (1984) addresses herself to teachers who wish to help their pupils improve their problem-solving skills. She uses the same model of problem-solving (entry, attack, review) as John Mason, but adds an extra phase, extension. Leone Burton also provides sets of what she calls organising questions and procedures which are linked to the phases and to each other as well as a list of skills. She describes and illustrates each of these, and thus makes explicit the many elements which go to making up problem-solving behaviour. Teachers may well find that lists of these questions, procedures and skills on the classroom wall would be helpful to remind everyone of these crucial ideas. The procedures for the phase of entry, i.e. getting going on a problem, include: explore the problem, make and test guesses, define terms and relationships, organise the information, etc.

These are all helpful ideas for getting into a problem, and can, if that is appropriate, be turned into questions, e.g. 'Have you made a guess and tested it?' or 'What information does the problem give you?'

The illustrative examples in the book are mainly of children's work, and a set of 30 problems, together with guidance to teachers for presenting the problems, is included.

Other resources for investigational work

The two classroom packs which the Shell Centre for Mathematical Education and the JMB have jointly produced are mentioned above, and the pack on graphical work is discussed. *Problems with Patterns and Numbers* was the first of these packs, and contains materials which are particularly appropriate for generating investigational work. The style and format are very similar in both packs, with worksheets, discussion points, advice and suggestions on classroom organisation, micro programs, transcripts etc. being provided.

Books and journals

A resource book for teachers with a very different style from anything so far discussed is *Maths in Context* (1986), a book of 'structured investigations' which were developed by the Modular Mathematics Organisation for use in Scottish Standard Grade Courses. Most of the activities are rather too directed to be classified as open-ended investigations, so this book is of limited value. However, as a halfway house for pupils (and teachers) who are very new to open-ended ways of working, it could provide a useful and reassuring starting point.

It is impossible here to do justice to the many books, journals and items of computer software, in addition to those already cited, which are a potential resource for investigational and problem-solving work. Several more are mentioned below, and others can be found in the Resources list. Some resources include suggestions on organisation; others are primarily collections of ideas to get children started.

Starting Points is the title of a book first published in 1972 and recently republished. This was one of the earliest books on investigating in the classroom and, in addition to being as fresh and lively now as it was then, serves as a reminder that some teachers have been using these approaches for some while now! As the title suggests, it is a collection of starting points, together with discussion of organisation and development.

Marion Bird's *Generating Mathematical Activity in the Classroom* contains a wealth of material from her own classrooms. Children's work abounds (most of the book is taken up with this) and she also provides valuable notes on setting up the activities.

Brian Bolt's *Mathematical Activities* (1982) and *More Mathematical Activities* (1985) are just that – lots and lots of interesting activities, clearly presented and with very little discussion.

Yet another useful book is *A Way with Maths* by Nigel Langdon and Charles Snape (1984). This is beautifully illustrated and written in a style and with language which make it accessible to relative novices - such as the pupils themselves. This is a book I would heartily recommend, both to teachers for their own use and as a welcome addition to the classroom library.

The two main professional associations for mathematics teachers both publish journals and other resources. *Points of Departure 1* and *Points of Departure 2* from the Association of

Teachers of Mathematics contain a multiplicity of short 'starters' which are particularly relevant to the area of the curriculum being discussed here. The ATM journals are *Mathematics Teaching* and *Micromath*, and the Mathematical Association publishes *Mathematics in school*. All these regularly include accounts of investigations in the classroom, thus providing a discussion forum for teachers, and helping to generate and disseminate new ideas – for this and other areas of the mathematics curriculum. Many of the teachers who are active in these associations have been at the forefront of the development of the new ways of working now incorporated into GCSE.

PRACTICAL WORK IN MATHEMATICS

Both the Cockcroft Report (DES, 1982) and *Mathematics from 5 to 16* (DES, 1985d) stress the importance of practical work. Practical work, according to *Mathematics from 5 to 16* is of three main kinds.

(a) 'There is the practical work which enables pupils to understand mathematical concepts.' For example, throwing dice can help pupils develop an understanding of probability.
(b) 'There is the practical work of measurement which needs to be done with a particular purpose in mind.'
(c) 'The activity itself might be conducive to a practical approach.'

Examples of the last would be carrying out the polyominoids investigation with the aid of Centicubes, or the hexomino investigations with squared paper and scissors.

Requirements

In line with assessment objective 3.17, each of the examining groups includes practical work in its coursework requirements, but the form this takes varies.

SEG provides a detailed example of a practical task and a list of suitable areas of enquiry for practical work. These include: modelling – polyhedra etc., finding heights of buildings, tessellations and patterns, and many more.

LEAG's notion of practical work seems somewhat idiosyncratic. It offers three sample coursework tasks which are labelled 'practical' but which do not fit any of the categories laid out above. The problems or situations to explore are indeed practical in the sense that they could have practical relevance in 'real life'. However, there does not seem to be much need for practical, in the sense of 'hands-on', work. The problems could be tackled with the aid of nothing more practical than a diagram – for example, one which involves working out a timetable for the mathematics staff in a school given certain constraints. Although these are interesting tasks, I would query whether they are what the national criteria, following the recommendations of the Cockcroft Report and *Mathematics from 5 to 16* intended practical work to be.

MEG has practical geometry – examples being simple surveying; model-making and packaging. NEA expects pupils to carry out practical work in which marks are awarded for 'skills in using instruments and equipment', and WJEC's examples of 'practical investigation' also involve pupils in making and measuring. Only LEAG's examples stand out as atypical. However, LEAG specifies that candidates may submit coursework tasks of their own choice, and it would seem very reasonable for pupils to offer, as alternatives, the kind of practical task which is clearly acceptable to all the other examining groups.

Why practical mathematics?

The case for practical work in mathematics, of all the kinds listed above, has been made so thoroughly in so many publications and over so many years that it is not necessary to go through all the arguments again in this book. Every mathematics education student has met the work and ideas of, for example, Piaget, Dienes and Bruner, and is aware that most learners need practical experiences in order to make sense of fundamental mathematical concepts and make these their own.

Suffice it to say that despite the fact that the crucial importance of practical work has been long recognised, this recognition has not been adequately reflected in public examinations at 16+. The inevitable result of this is that practical work has been neglected in most classrooms, primary and secondary, but especially secondary. As with oral work, the justification for practical work lies in its power to improve the learning experience of pupils, rather than in whether or not it is formally assessed. However, the inclusion of practical work in GCSE assessment legitimates it for pupils of all abilities and not just, as has so often been the case, for less-able pupils.

Implications for classroom organisation

How must we organise mathematics teaching to allow ample scope for practical work of all kinds?

Equipment

First and foremost, suitable equipment must be readily available. Much of this equipment is very simple and not expensive, but there must be enough to go around so that each classroom has its own store to which pupils have easy access. Obviously, basic geometrical equipment is needed – rulers, pencils, protractors, pairs of compasses, pairs of scissors, measuring tapes, etc. Then there is a variety of papers – squared, graph, plain, spotty (square), spotty (triangular), isometric, tracing, gummed. Large sheets as well as A4 size would be useful. Cheap coloured paper is useful for mounting (e.g. sugar paper), especially for large displays, and smaller sheets of plain coloured or black paper for mounting smaller pieces of work. You will also need thin card (for model-making), occasionally heavier card, glue and Sellotape, a long-armed stapler and mounting pins. Other equipment for practical work includes pin-boards (Geo-boards), elastic bands, straws and other constructional materials, mirrors, Centicubes, models of other solids (or get the children to make them!) and so on. Don't forget junk materials – empty boxes and cartons, for example. In addition to packaging projects, there is lots of scope for work on volume and area, on two- and three-dimensional tessellations, on design, on scale drawing, enlargements, ratio ... Every classroom should have a box of junk materials which anyone can use or add to, and a box of scrap paper for trying out ideas (written or practical).

Classroom space and furniture

It has been a source of great frustration for many mathematics teachers that our subject has been seen as theoretical and desk-bound rather than practical. Unlike science and craft teachers, whose need for laboratory and workshop space as well as plenty of space for storage of equipment has been unquestioned, mathematicians have had to justify even their need to be based in specialist rooms. Anyone who has had experience of teaching in four or five different classrooms in the course of the week knows how difficult it is to organise practical work, especially if those rooms are used for other subjects the rest of the week. The inclusion of practical work in GCSE coursework means that at last we have unassailable needs and rights, in common with colleagues in other subjects, for our classrooms to be organised as workshop areas with appropriate equipment and storage space.

Ideally, these rooms should be furnished with tables rather than desks. This not only makes it easier to do practical work where a large, flat surface is quite often needed; it also makes it easier to organise groups for investigational work and oral work. If you are stuck with desks for the time being, you might consider rearranging them so that pupils sit facing each other across the desks in small groups. Perhaps you could acquire one table in addition which could be used whenever someone has a particular need for a flat surface. These latter suggestions however, depend on at least having a classroom which is used primarily for mathematics – this really is crucial. Teaching mathematics 'on the fly' is only feasible if it is desk and text-book bound – and GCSE mathematics is neither of these!

There are several examples in the next chapter of work of a practical nature. It will be apparent to readers that this work would be much more easily undertaken in conditions such as are described above than it would be otherwise. It is essential, therefore, that mathematics teachers press their case for properly equipped mathematics teaching areas to headteachers, colleagues and the local authority, and continue to do this until mathematics is recognised as being a practical subject.

In the meantime, however, and notwithstanding this strong

plea for appropriate teaching conditions, it is possible for teachers and children to gain experience of practical work – albeit in a more limited way – in less than ideal circumstances. It means that everyone will have to be working on a similar project so that the variety of equipment needed is reduced, and the equipment specifically needed will if necessary have to be carried around the school to wherever it is required. This is not to be recommended in the long term, but would still be better than not doing any practical work at all.

Extended pieces of work

Nothing has yet been said about extended pieces of work, simply because regarding this as a separate category creates a false distinction. Any piece of work on which a pupil has spent a substantial amount of time (that is, more than a few lessons), and which has involved several stages in its development, can be considered an extended piece of work. Both practical work and investigational work often (but not always) lend themselves to development over time. An investigation could be 'completed' in fifteen minutes, or a child might spend a week or more of class lessons and homework time on it, looking at different aspects, refining generalisations or trying out related activities.

Several of the classroom examples in Chapter 4 are of extended pieces of work, some practical and some investigational. As Frances Bestley points out in her account, the same 'starter' may have very different outcomes in terms of time devoted to it by different pupils. One or two children may be sufficiently intrigued to want to pursue the activity for much longer than everyone else or to take it in different directions.

The microcomputer, too, may serve to generate the kind of enthusiasm which leads to pupils pursuing a particular line of work for a lengthy period. Programming projects, especially turtle graphics in LOGO, often have this effect, as students are motivated by partial successes to wish to produce the graphics of their choice. On the way, much is learned – mathematical ideas, programming skills, problem-solving skills – and the end product can be admired by all.

Examples of extended pieces of work arising out of the use of a microcomputer can be seen in the recently published *Secondary Mathematics with Micros In-Service Pack* edited by Jo Waddingham and Alan Wigley (1986). The pack includes micro programs, examples of pupils' work and much discussion and suggestions for classroom activities.

CONCLUSIONS

So although mathematics teachers do not, in general, need new textbooks for GCSE, many additional resources are indeed needed if teachers are to be enabled to develop the new aspects satisfactorily – stable mathematics classrooms which are well equipped with materials for practical work, plenty of calculators, one or more microcomputers permanently available, and a mini-library of books which can serve to stimulate pupils' and teachers' thinking about projects and investigations.

Also needed is a substantial mathematics staff library of resource books and packs, some of which have been cited in this chapter. The Resource list at the end of the book offers a wide range of available materials, some of which would be suitable for pupils to use on their own; others which are more appropriate as a teacher resource.

Of equal importance is time – teachers need time for collaborative work with colleagues if they are to develop and reflect on the new skills which are now needed. Every school with several mathematics teachers has a built-in resource for inservice work. This is because possibly the most effective way of developing new skills is to try things out with colleagues and discuss with them what is happening. And teachers in small schools need time to meet with other local teachers. Much can be learned, about organisation and assessment, through several teachers experimenting with the same project in their own classes and then sharing experiences.

Another way forward is through one or other of the many inservice courses and other activities for mathematics teachers available in different parts of the country. There are award-bearing courses (diplomas and higher degrees) offered by universities, polytechnics and colleges. These institutions also often provide short courses which do not lead to any qualification. The local authorities also provide a range of short or longer courses, and the professional associations run conferences, workshops and teachers' meetings. All these activities, in their varied ways, offer teachers a forum for the discussion of issues as well as opportunities to share ideas with colleagues in other schools, and explore and critically evaluate new material for the classroom.

GCSE offers really exciting opportunities to work in the classroom in ways which, although new to most teachers, have long been recognised as stimulating and enriching – both of mathematical skills and of pupils' self-esteem and confidence.

4 Classroom Examples

> 'Lessons teach you to do sums, and things of that sort.'
>
> *Lewis Carroll*

INTRODUCTION

A number of teachers who are already working with their pupils in the new ways discussed in Chapter 3 provide accounts in this chapter of classroom activities together with examples of their pupils' work.

The majority of these teachers work in schools using the SMILE or OMEGA systems, where investigational and practical activities and collaborative styles of working play an important part in everyday classroom life. As a result, teachers working within these schemes have considerable experience of what are, for most people, very new areas of mathematical education.

As stated in Chapter 3, SMILE is a curriculum development project which uses resource-based learning. Typically, in a SMILE classroom pupils will be working (individually or in small groups) on a range of topics at any one time. The SMILE bank of materials, developed by teachers over a period of years, includes a range of investigational activities, some of these being microcomputer programs (see Microsmile in Resources list).

OMEGA is organised as a modular scheme and draws on material from a number of sources, in particular SMILE and SMP. In an OMEGA classroom, pupils are likely to be working on the same topic, but perhaps using examples differentiated by difficulty.

The examples in this chapter include: extended class projects, both practical and investigational (statistics, Christmas designs and Chinese triangles); investigations carried out by individual children; class lessons on a practical or applied topic (planning a kitchen, Winston's garage). Several of the teachers, in addition to describing the activities, have explained why they find these ways of working exciting and rewarding.

In the previous chapter the various aspects of the new ways of working are considered more or less independently for the purposes of clarification. Classroom reality, of course, is very different. Although one may introduce a particular activity primarily, perhaps, to develop a particular aspect, e.g. practical work or oral work, other elements will be involved. An extended project may give a child experience of, for example, collaborative work, oral work and practical work. An investigation or a LOGO project might provide opportunities for collaborative work or might be tackled individually – in both cases, explaining the work after completion would provide opportunities for oral work, and so on.

As it happened, none of the examples teachers were able to provide for this book was of children's work around the micro. In Chapter 3, however, there is a discussion of the role of the microcomputer in developing and extending these new ways of working, and readers may also find the many accounts in professional journals and magazines of use to stimulate and generate ideas.

The final section of the chapter draws together the teachers' accounts, picks out common themes and offers suggestions for implementing some of these projects in more traditionally organised classrooms.

CLASS PROJECTS

Bridget Perkins: A statistics project in a girls' comprehensive school

The activity described below was the conclusion to a module of work on 'representing data' which is part of the first year OMEGA scheme. The first-year pupils are taught in mixed-ability form groups. The girls in my group are very friendly and cooperate well with one another.

The work we had been doing recently covered very basic statistical representation, so the class had met pie charts, bar charts and pictograms. The activity was introduced by looking at some opinion polls which I had collected from a variety of

▼ *Fig. 1. The original size of this work was 1m × 60cm.*

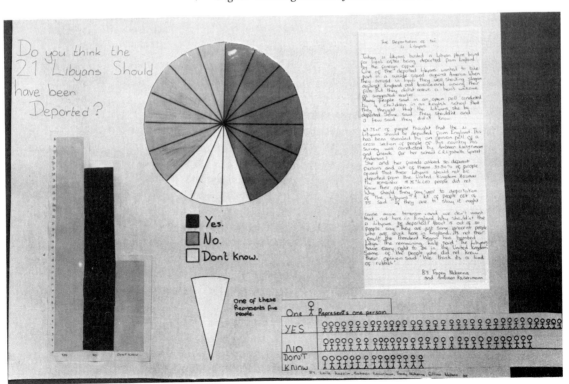

newspapers. We discussed the sorts of question used in polls and the sorts of representation that were effective.

Organisation

The class was then put into groups. I tend to organise the groups myself and change the groups for each new piece of work. This is so that they experience cooperating with different people. I try to keep a range of ability in each group.

Activity

The idea of the activity was that each group was to carry out a survey and present its data as though for a newspaper. The children first chose the question (I did insist that this should be related to a current news item). They then conducted the survey for homework – over the weekend – so that the poll sample was from outside school. The next five mathematics lessons were involved with presenting the data and writing a newspaper article to go with it. Dealing with real data led to several problems which had not been touched on in the representing data module. For example, the class were used to nicely-divided pie charts – a sample size of 85 presented some problems to a class that had yet to cover angles.

All through the activity I was very strict about the type of question that I answered. I wanted to be in the classroom as a resource and not as a decision-maker. Questions like 'How do we use an angle indicator?' got answers, whereas 'How do we show this?' did not. The girls found this very difficult to cope with at first, but they did become used to it.

▼ *Fig. 2. The original size of this work was 1m × 60cm.*

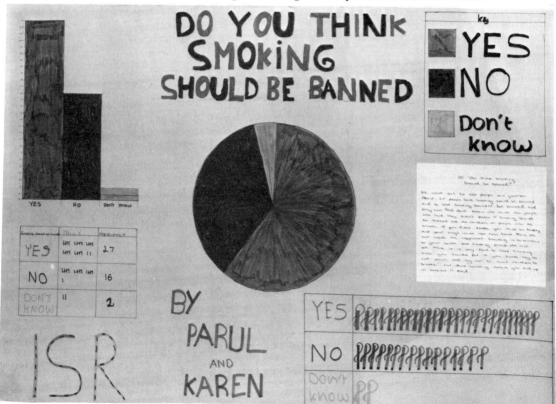

Work

A lot of mathematics came out of the activity, apart from the simple arithmetic and statistics involved in manipulating the data. The production of posters involved measuring and accurate drawing. Writing the article required an understanding of the data and a judgement as to how it could be clearly presented.

A discussion followed the completion of the posters. Each group criticised their own poster and then other groups' posters. The discussions that led on from this about bias in statistics got quite complex. However, the girls found it much easier to follow than an abstract discussion of the same topic, as they had all experienced the problems involved.

On the whole I felt it was a successful activity. The girls got involved and were happy to spend two weeks on the project.

Bridget Perkins: A Christmas design project in the same school

The activity was carried out by a third-year class. The class have been following SMILE since they were first years. The group is a middle set, unchanged since the first year. The setting has

▼ *Fig. 3*

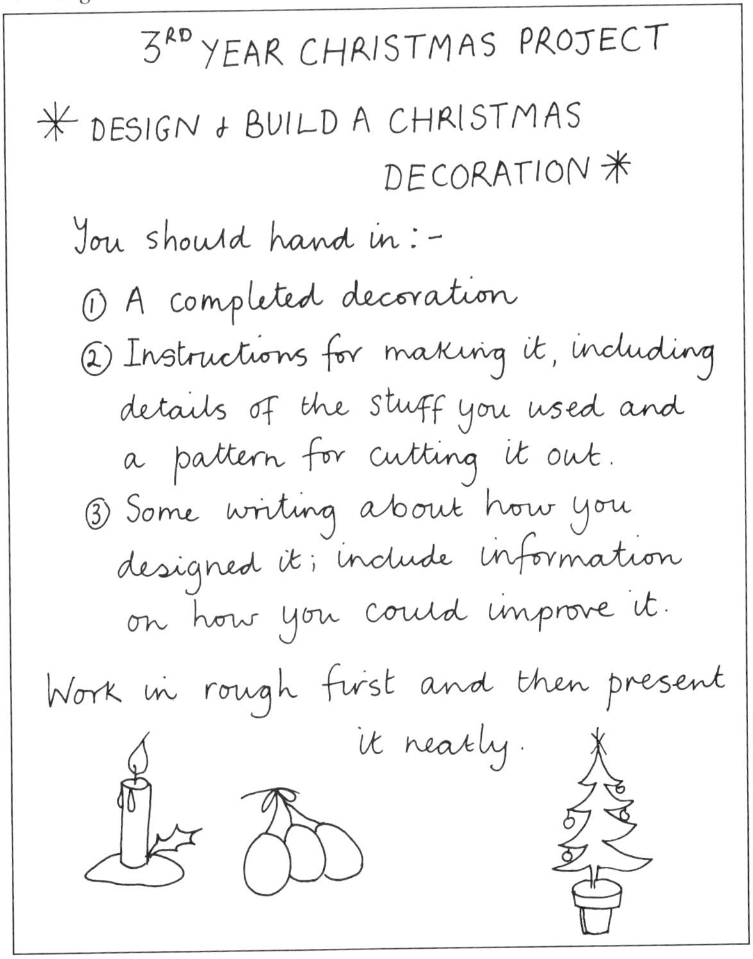

> 3RD YEAR CHRISTMAS PROJECT
>
> ✶ DESIGN & BUILD A CHRISTMAS DECORATION ✶
>
> You should hand in :-
>
> ① A completed decoration
> ② Instructions for making it, including details of the stuff you used and a pattern for cutting it out.
> ③ Some writing about how you designed it; include information on how you could improve it.
>
> Work in rough first and then present it neatly.

▼ *Fig. 4*

C.W.K.

SNOW FLAKE MOBILE

For a snow flake mobile you will need:
White card,
string,
coloured card and
Ribbon.

1. Collect four squares of white card 8cm² and fold twice like this.

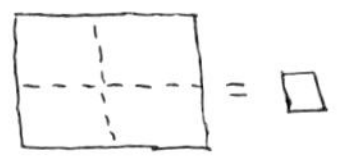

2. Cut shapes round it like this, then open up and you should have a snow flake shape.

3. Cut two strips from the coloured card 20cm by 4cm. Get the string and hang it from each end to look like this. — string.

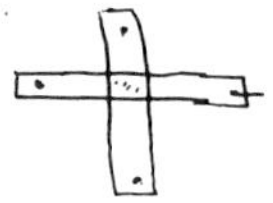

4. Attach the snow flakes to the end of the string and to finish put the ribbon on the top so you can hang it up.

HOW TO IMPROVE IT

You can improve it by adding glitter on the flakes and I could have put some holly on aswell. A good place to hang it would be in a doorway, because it would look nice there and blow whenever anyone walked in and out.

By Helen Adah 3KR

	been gradually eroded and the ability range is now fairly wide. The group does not work well as a class, since they are taught in form groups in every other subject.
Stimulus	The activity did not really require a stimulus – to mention Christmas decorations was enough. A worksheet (Figure 3) was handed out simply to give some structure to the work and, by setting objectives, to prevent girls spending a week making paper chains.
Organisation	I did not set up any groups – some girls worked alone, some as friendship groups. I do not think this is ideal, but I allowed it because it was the end of term.
Activity	The idea was to turn making Christmas decorations into a design project. This involved writing instructions, lists of materials and equipment, as well as costing and producing a finished article. I was worried because this activity is centred around a Christian festival and so is not multicultural in itself. However, I envisaged it as part of an overall multicultural mathematics curriculum. I misjudged the time required – several groups did not get their work written up in one week – and of course it would seem ridiculous to continue after Christmas.
Work	I did want to encourage originality and, although some groups did copy known designs, on the whole I was pleased. I felt the most important parts of the activity were:

1. Following through a procedure of planning, building, costing and criticising their work
2. Being able to write clear and precise instructions for someone else to follow.
3. The use of mathematical skills such as measuring, accurate drawing, using symmetry, nets of solids etc.

An example of pupils' work is shown in Figure 3.

Martin Marsh: Chinese triangles in a boys' comprehensive school

The class	I decided to use the Chinese/Pascal's triangle investigation that I had seen in *Investigator* 6 with my class of second-year boys (second of four sets) who had been doing work of this type for about a year and were now quite confident.

Introduction of the activity

I began the lesson by putting on an overhead projector the first three lines of the triangle:

```
        1
      1   1
    1   2   1
```

I then asked the class what the next line might be. Although I did not get the answer I wanted, the children were asked to justify their guesses or ideas, and this was a useful exercise. The next line, 1 3 3 1, was a big clue, and the pattern was soon established.

The children were given the opportunity to write down some more lines and note anything of interest. I gave them hexagonal paper to help them spot patterns, and very soon they realised that the whole triangle was symmetrical. Some boys started putting sequences along diagonals; one added up the numbers on each row and obtained the sequence 1, 2, 4, 8, 16, ... His friend pointed out that this was 2, 2^2, 2^3, 2^4, ... but could not see how 1 fitted into this pattern.

Development

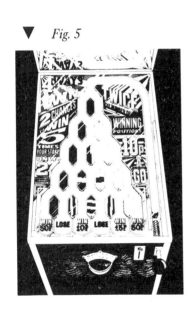

▼ *Fig. 5*

At the beginning of the double lesson the next day, some of the children's ideas were put on the overhead projector for everyone to share. This gave rise to a certain amount of discussion about how their work could be extended. I posed two further questions:

1. What if we built up these triangles using modulo arithmetic?
2. Is there any connection between this triangle and the poster of the pinball machine that I put on the notice-board a few days ago? (Figure 5)

The first question led to a considerable amount of exploration of pattern which most of the boys got involved in immediately. Some decided to ignore by questions and pose their own, for example:

1. What if you start

2. What happens if we multiply, subtract or divide to get successive rows instead of adding?

e.g.
```
           1                          2
          1 1                       2   2
         1 0 1                     2  4  2
        1 1 1 1                   2  8  8  2
       1 0 0 0 1                 2 16 64 16 2
```

(Some wondered why this pattern was the same as the modulo 2 Chinese triangle)

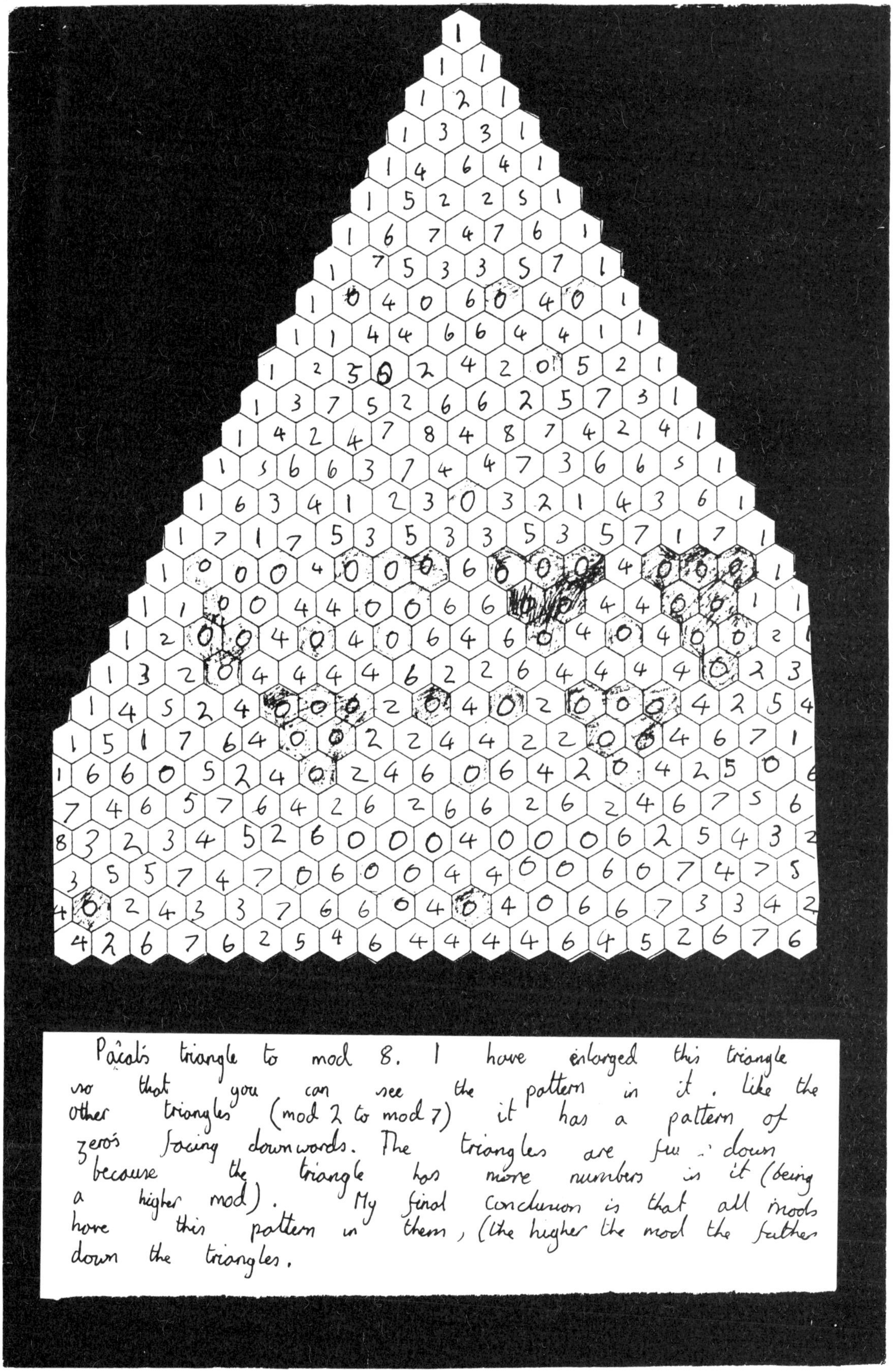

Pascal's triangle to mod 8. I have enlarged this triangle so that you can see the pattern in it. Like the other triangles (mod 2 to mod 7) it has a pattern of zero's facing downwards. The triangles are further down because the triangle has more numbers in it (being a higher mod). My final conclusion is that all mods have this pattern in them, (the higher the mod the further down the triangles.

▲ *Fig. 6*

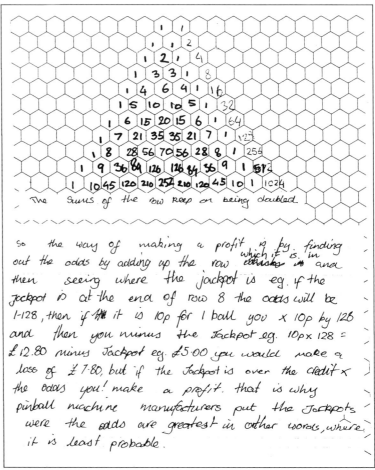

▲ Fig. 7

3. What other patterns can we find in the standard Chinese/Pascal's triangle? (Patterns to be found include the Fibonacci numbers, powers of 11 and 2, the counting numbers, the triangle numbers, and so on.) An example of work is shown in Figure 6.

Only a few boys were interested in the pinball machine problem, so I put them into a group and we discussed some simple probability ideas. On a simplified diagram of the pinball machine (just two levels) we talked about the number of different routes the ball could take and how they were all equally likely, but that two routes ended up in the same place. They then investigated a more complicated pinball machine (three levels) and noticed that there were eight routes, but that three ended in one place and three in another. It was not a big jump to notice the relationship between this and the work they had been doing on Pascal's triangle, but they did check they were right by looking at the 1 4 6 4 1 row. I then asked them where they would put 'win' on their own machines to ensure the machine made a profit. After this they investigated much more complicated machines. Samples of work are shown in Figures 7, 8a and b.

Conclusions

The whole investigation kept everyone happily working for a single lesson, two double lessons and three sets of homework. The final homework was to write up their results and present their work. I discussed each child's work with him, and it was clear from many comments that there was still plenty of scope to extend this work. Nearly all the children had ideas for extensions. To provide an opportunity for this I said that they

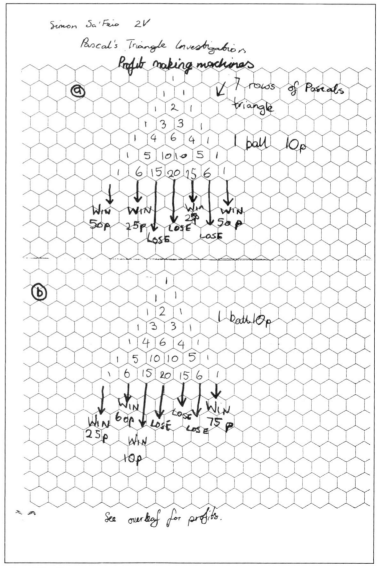

▲ Fig. 8a.

could submit a completed extended piece of work as coursework to count as part of their summer examination mark. Everybody did this and I allocated 20 per cent of their overall mark to this project. (I found it a very useful exercise, also, in assessment of work of this kind.)

The boys got an enormous amount of mathematics out of the Chinese triangles investigation, and the quantity of work they produced was phenomenal. There was also pleasure in 'discovering' patterns and enjoyment in sharing their 'discoveries' with their friends and teacher. For me, Chinese triangles is a first-rate open-ended piece of investigational work.

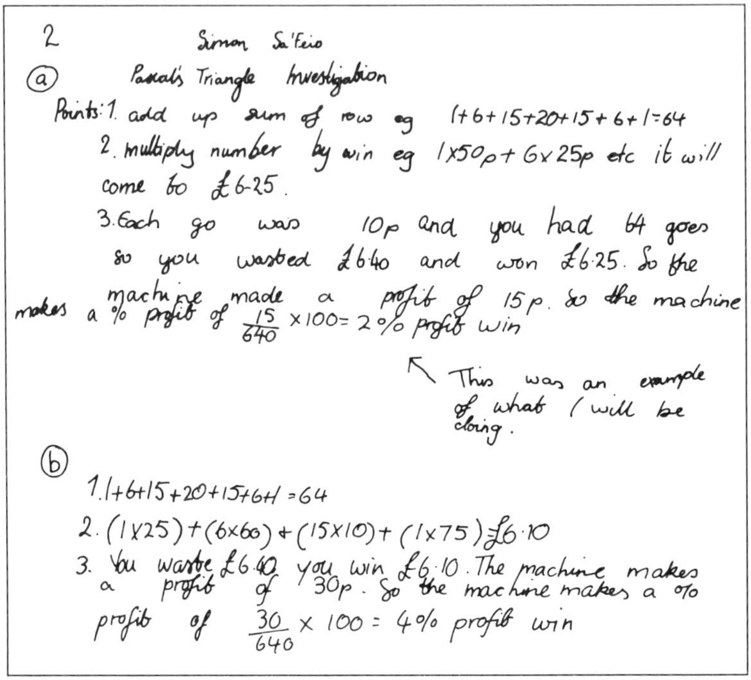
2 Simon Sa'Feio

(a) Pascal's Triangle Investigation

Points: 1. add up sum of row eg 1+6+15+20+15+6+1=64

2. multiply number by win eg 1×50p+6×25p etc it will come to £6·25.

3. Each go was 10p and you had 64 goes so you wasted £6·40 and won £6·25. So the machine made a profit of 15p. So the machine makes a % profit of $\frac{15}{640} \times 100 = 2\%$ profit win

This was an example of what I will be doing.

(b)

1. 1+6+15+20+15+6+1 = 64

2. (1×25)+(6×60)+(15×10)+(1×75) £6·10

3. You waste £6·40 you win £6·10. The machine makes a profit of 30p. So the machine makes a % profit of $\frac{30}{640} \times 100 = 4\%$ profit win

▲ *Fig. 8b*

CLASS LESSONS

Tony Purcell: problem-solving in a girls' comprehensive school

Problem-solving plays an important part in the mathematics curriculum in my school. Both 'pure' and 'practical' tasks are undertaken by pupils, and the tasks are mainly of an investigative or experimental nature. Work in the lower school is organised into modules (we use the OMEGA scheme) and tasks are chosen to be appropriate to the module. This avoids tasks being seen as just a one-off lesson.

The sort of lesson which starts 'Today we are going to study the addition of fractions ... Here is how you do it ... Now turn to page whatever in your textbook and get on with it' is very rare. Collaborative learning is emphasised, and teachers often teach by assisting or taking part in work with individuals or small groups.

However, there are still problems. Pupils tend to ask 'Where do I find the booklet?', 'What do I do here?' – the latter before they have even read the task. Pupils are still frightened to have a go. You will notice that the work reproduced here is 'fair copy' – the working-out and notes have been 'lost'. This is partly, I suspect, as a result of the natural tendency of pupils to want to present neat work to the teacher and partly as a result of the school ethos which encourages neatness. The emphasis in the department, however, is on communication, both verbal and written, and not on neatness.

It is fair to say that many pupils would prefer to do exercises from text books – as long as they were not too taxing and they could get lots of red ticks. So why are we so keen to work in this way?

First, it has improved the attitude to mathematics, particularly in the lower school. Pupils enjoy the subject, although to be fair, if teachers are not careful, in some classes pupils can get away with contributing little. However every pupil at some stage suddenly shines at something and, gradually, as a result, her confidence will grow. The immediate result has been a vast improvement over recent years in examination success among the 'less-able' and 'average' pupils – we have still some way to go at the upper end. More importantly, there has been a change in attitude. Pupils are more confident about mathematics and less worried if they do not remember a method. Pupils tend to develop their own methods, and although this takes more time, they do get there in the end. This has produced some problems in other subject areas, but on the whole once teachers become more flexible, they are surprised at the extent to which pupils can solve the mathematical problems which come up in other subject areas.

Second, it produces a greater understanding of individual pupils' needs amongst teachers.

Third, socialisation in the mathematics classroom is improved. Pupils tend to want to work together.

Fourth, pupils are much more confident with equipment, can verbalise better and, slowly but surely, improve their numeracy and their spatial skills – permanently.

Last, on the converse side, classrooms tend to be noisier, appear more chaotic and are certainly more draining on the teacher. Preparation is harder, and much more effort goes into the job. Also, I would have to admit that pupils have regressed from 'the good old days', since it is clear few could attempt a long-division sum without the aid of a calculator and no one knows what a log book looks like.

Winston's garage

This was attempted by most third-year pupils as part of a logic and planning package. The work is based on a GAIM worksheet, which is reproduced here (Figure 9). (Author's note: see Chapter 5 for description of GAIM). The organisation was as follows:

1. The pupils were given the worksheet and a daily log sheet.
2. They were given a 45-minute lesson to consider and plan their response, and a homework to complete the log sheet.
3. They were told by the teacher that there was more than one way of doing this and that they were to try to find more than one way. Also, the teacher asked for written explanations as to why the pupils had chosen a particular solution. There was no other contribution by the teacher. The pupils discussed the work among themselves, but the 'final' version was not strictly collaborative because of the homework element.
4. The pupil whose work is reproduced (Figures 10a and b) is of 'average' ability.

▲ *Fig. 9*

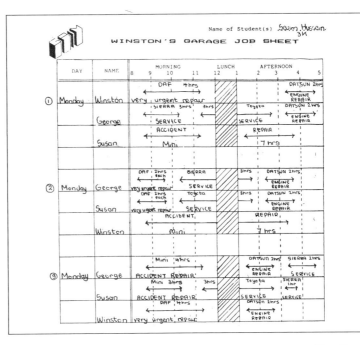

▲ Fig. 10a   ▼ Fig. 10b

① Winston is the boss, he should do the less work. He does the daf because it's a very urgent repair and so his experienced he would be able to finish it quickly.
Susan does the longest repair but she has an hour off because she will get more tibled then Geogre or Winston would, or she could have one hour more for Lunch.
George does three different Jobs because he would be more determened to learn more things and it would keep him busy.

② For this one they have all the same hours. Susan may be against having less hours because she's a woman. So Winston the boss would make sure Susan and Geogre would have the same amount of hours and himself. Maybe Winston might think it's wrong for the boss to have less hours so he would have the same amount of hours has his workers to set an example.
[him-sielf]

③ Winston has made sure they all work toghter this time. He might think it's good for him and his workers to work toghter. They get to know each other better and this makes them trust each other.

87

▼ *Fig. 11a*

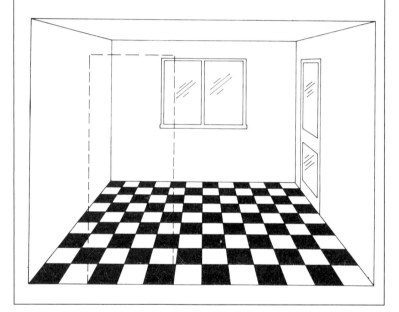

Monica Ramsay: Planning a kitchen in a girls' comprehensive school

The 'planning a kitchen' activity was carried out by third-year pupils who had never worked in this way before. It took about two lessons, and a lot of oral work came out of it as well as the finished plans.

The girls were provided with SMILE card 1310A (Figure 10a), the kitchen plan and cut-out sheets (SMILE 1310D, 1310E and 1310F). The cut-out sheets have much more furniture than can be used, so the girls had to choose what they needed. They also had to do scale drawings for their finished work, as the plan and cut-outs are on a larger scale.

▶ *Fig. 11b*

① Have you left enough for a person to work in your kitchen? Yes we have left enough room.

② Can both doors be opened safely? Yes the doors can open safely

③ Which items have you left out? We have left out the deep freeze, a single sink unit, a round table, a twin tub washing machine, refrigerator, a cupboard with working surface because we have got too many cupboards with working surfaces, and we don't need a deep freeze because we have got a fridge freezer, and we didn't need the spin dryer because we have got an automatic washing machine, and we don't need a single sink unit because we have got a double sink unit because it is useful.

④ Can all the cupboard doors be opened? Yes all the cupboards can be opened

⑤ Is the cooker near the door, if so is this a good idea? No the cooker is not near the door.

⑥ Is there somewhere for people to sit and eat? in your kitchen?

⑦ Where should a builder put electricly sockets? he or she should put them near the hall door a cooker socket should be put near the cooker and the water supply should be put near the sink.

▼ *Fig. 11c*

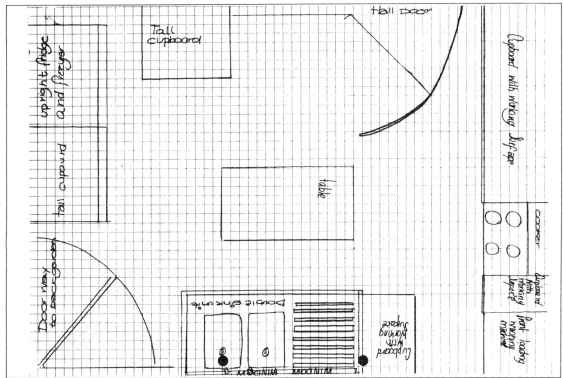

INDIVIDUAL WORK

Frances Bestley: Investigations in a mixed comprehensive school

I have taught using investigations ever since I started teaching five years ago. As I was a probationer when I started and all methods of teaching were difficult, I think it was easier to come to terms with investigations and to accept that sometimes they would be excellent and sometimes disastrous.

My school is a mixed comprehensive school with more boys than girls (about 70 per cent of the pupils are boys). We use the SMILE system throughout and, as all our classes are mixed-ability, most of the students are used to a non-traditional way of working. Most mathematics classrooms are arranged so that the students can work in groups.

When I do an 'investigation', I tend at the moment to do it with the whole class. There is no particular advantage in this, although as I use an individualised scheme I think it is good to have the opportunity to have the students working in a larger group and to discuss mathematics with them as a class.

I generally introduce the starting point from the board by showing a few examples and then getting the students to explain what is happening and probably do some examples themselves on the board. We will also generally have a discussion about what is allowed. For example, in a polyominoes investigation which required students to find out how many tetrominoes and pentominoes they could find, a second-year class decided that they would not allow rotations and reflections to count as different, whereas a first-year class decided that they would allow this. (Author's note – see the description of polyominoes in Chapter 3.) The students have to set their own questions and boundaries. In order to let the students be in control, I think it is important that I have not 'finished' the investigation myself.

Students are then given as much time to do the work as they want. If someone is getting really stuck and feels they are getting nowhere, I may think of some things they could look at but whenever possible I try to let them work out what they are going to do. Some students will spend a double lesson (eighty minutes) and then decide they have had enough, and others will still be working on the same topic lesson after lesson. It depends very much on the student and the investigation.

In the lower school I tend to set the 'how many different ones are there' kind of investigation such as 'Polyominoes' or 'No circles'.[1] In the upper school I tend to set a more formal type of investigation and encourage the students to work in a 'logical' way, so that they can see how things are changing and pick up patterns more easily. It is quite remarkable to see how much the fourth-year class are getting out of investigations and how much more they enjoy them as they do more of them.

I think the most important thing about using investigations is to persevere. Some students do find it very difficult to start with, but enjoy it eventually. A fifth-year girl who lacked confidence in mathematics refused to do anything apart from investigations once she had discovered how good they were. Different classes

▼ Fig. 12a

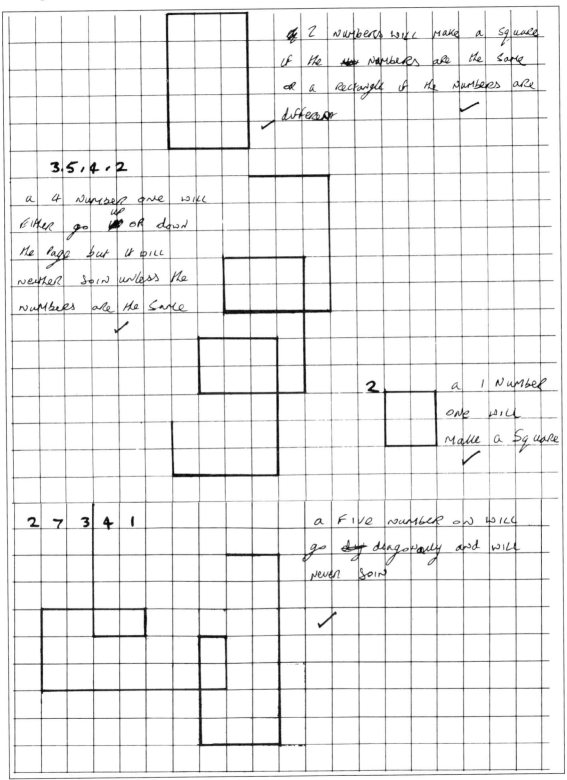

can also respond quite differently to the same investigation. I once gave exactly the same investigation to a fourth-year class and a second-year class: the fourth years struggled and the second years romped through it.

Investigations are an important part of the mathematics curriculum, and as such there will be an increasing commitment to do them well. It is threatening for some teachers, as they have to relinquish some of their control to their students. They have to be flexible, and there is no 'right answer' or even right way of doing an investigation. However, when an investigation is successful, it involves the students in active discovery of mathematics and greatly enhances their motivation.

[1]An account by Frances Bestley of a 'No circles' class activity and examples of her pupils' work is in *Splash* No. 111110, June 1986 (see Resources list).

Frances Bestley's pupils' work

Fourth year boys' work on spirals

The boys, one of whose work appears in Figure 12a and b, were introduced to spirals by means of a SMILE booklet, No. 0728, which takes the learner through the various stages of building up spirals and asks them to investigate the different patterns which emerge. Spirals could also be introduced as a class activity, for example, in the way outlined in *Investigator* No. 5 (Figure 13). The microcomputer program referred to is one of those on Microsmile 1 (see Resources list).

▼ *Fig. 12b*

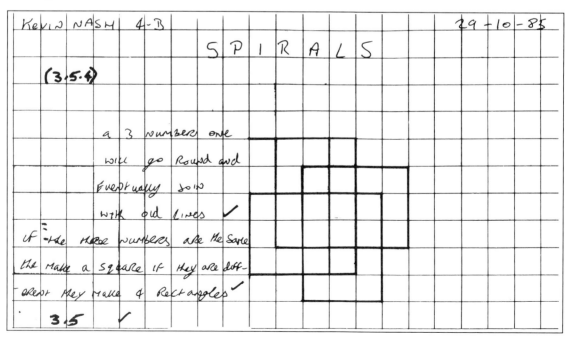

A spiralling investigation

The robot line painter only works on the lines of a square grid and can only make right turns when it arrives at corners. The line painter is programmable. For example, the program (1,3,2) makes the line painter go one unit from its starting point, then do a right turn, go three units, do another right turn and then go another two units. The line painter will continue to repeat programs until it arrives back at its starting position or its vanishes off the grid.

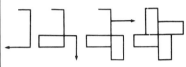

Line painter's path (1,3,2).

What happens with different programs? A few are shown below.

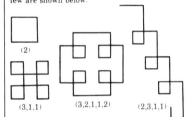

The description above can be used to introduce *A Spiralling Investigation*. If children

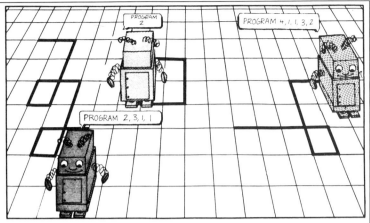

put themselves in the place of the line painter it helps them to sort out left from right more easily. Drawing some of the line painter's possible paths using an overhead projector helps get the rules across.

A Spiralling Investigation allows children across a range of ability to work at their own level. There is much to discover and much enjoyment to be had plotting the path. For some, it will be an exercise in drawing and in following a system. For others it will be about classifying: choosing families of spirals and finding out about the similarities and differences. It may prove to be an opportunity to explore using a microcomputer[1] (and this takes the investigation a whole lot further). *A Spiralling Investigation* could be extended to using algebra to describe what is happening, exploring negatives, negative negatives, zeros... *A Spiralling Investigation* should throw up plenty of questions. Below are a few that might be posed:

- Some spirals are closed; others go on forever. Which? Why?
- How big are the loopy bits? Why?
- Some have holes in the middle. Some have loopy bits. Which? Why?
- What difference does using isometric paper make?
- Which sets of numbers give identical spirals? Similar spirals?

[1]. A utility program SPIRALS can be used for further exploration using a microcomputer. For further details see the Frogs footnote below.

▲ *Fig. 13*

A fifth year boy's work on centuries

The 'starter' for this investigation was SMILE card No. 0700. This asks students to investigate ways of making 100 using all nine digits in order – and in reverse order too. A pupil's work is reproduced in Figure 13.

COMMENTS

The unifying thread in these otherwise diverse accounts is the commitment of all the teachers to ways of working which open mathematics up to investigation and exploration. The pupils' responses demonstrate their interest and involvement, and for these children, clearly, the mathematics they do is meaningful and personally relevant.

There are many ideas which readers could try out in their own classrooms. There would be no need to reorganise a 'traditional' classroom to any extent. Several of the examples, such as Chinese triangles, Winston's garage and planning a kitchen were introduced as a class activity. You might wish to rearrange a few desks so as to make group work and discussion easier – nothing more radical is required.

Some of the ideas could be adapted for your particular circumstances or to fit in with a local or topical event. The

▲ *Fig. 14*

statistics project and the design project especially lend themselves to this. For example, you might develop a project around a forthcoming event, such as a school fête. Pupils could be asked to design containers for the home-made sweets and novelties which are to be sold at the fête. Another class might carry out a survey of local opinion on a controversial proposal which will affect everyone's life. (Are there plans for a new motorway, a power station, a shopping complex, an airfield, school amalgamations?) This could then be written up and presented for publication – perhaps in the school magazine or a local newspaper. Statistics projects, particularly, are appropriate for children across a range of ages and abilities, as they can be tackled according to the sophistication and experience of the individuals concerned.

The investigation starting points provide many possibilities. These too could be introduced as class lessons (as in fact happened in several of these accounts), through a microcomputer program, or by means of a worksheet. Pupils could work individually or in small groups. Teacher-led or teacher-chaired class discussion could come out of this if wished. Martin Marsh explains how he used a class discussion to pool the children's ideas on the Chinese triangle investigations. The permutations are many and varied – there is no need to stick to just one pattern, and indeed many reasons to ring the changes.

Teachers often comment that children are reluctant to present 'untidy' work, and that working notes tend to disappear in favour of a final 'fair copy'. It is not easy to persuade children (and especially girls) that the thinking and exploration they do in the course of an investigation is worth keeping, and it is pleasing that one or two of the examples in this chapter show mathematics in the process of being made!

In Chapter 5, which deals with assessment, there are some additional examples of children's work on investigations. These were carried out under examination conditions. There are also suggestions for readers who wish to practice assessing investigational activities.

5 Assessment in GCSE Mathematics

PROPOSITION I. To find the value of a given Examiner.

Example. A takes in 10 books in the Final Examination, and gets a 3rd class; B takes in the Examiners, and gets a 2nd. Find the value of the Examiners in terms of books. Find also their value in terms in which no Examination is held.

Lewis Carroll

Principles

We believe that there are two fundamental principles which should govern any examination in mathematics. The first is that the examination papers and other methods of assessment which are used should be such that they enable candidates to demonstrate what they do know rather than what they do not know. The second is that the examinations should not undermine the confidence of those who attempt them. Because the syllabuses which will be prescribed and the papers which will be set will be the greatest single factor in influencing the mathematics teaching in secondary schools in the coming years, we believe it to be essential that the examination should provide suitable targets and reflect suitable curricula for all the candidates for whom the examination is intended; and that in order to achieve this it will be essential to provide a number of different papers so that candidates may attempt those papers which are appropriate to their level of attainment. (The Cockcroft Report, paragraph 521. (DES, 1982))

As is apparent from the discussion in Chapters 1 and 2, the views expressed here have greatly influenced the national criteria for GCSE and are reflected in the schemes of assessment offered by the various examining groups. The phrase 'enable candidates to demonstrate what they know, understand and can do', which echoes Cockcroft's 'enable candidates to demonstrate what they do know rather than what they do not know', is reiterated time and again in the official documents. The corollary of this requirement is assessment techniques which allow for differentiation between candidates of differing attainments: in other words, that now familiar phrase, 'differentiated assessment'. In the case of mathematics, this has been interpreted as both the provision of a range of examination papers which between them can cater for different levels of attainment, exactly as the Cockcroft Report recommended, and a coursework requirement which pupils can fulfil by carrying out work appropriate to their own individual capabilities.

WHY ASSESS?

Assessment and evaluation in educational contexts have many purposes and can take many forms. For example, there is diagnostic assessment, whose purpose is to provide information based on which a teaching programme can be devised. Informal school tests often have this function – a teacher may give a test to find out how much her pupils have understood or remembered of a topic, prior to preparing a scheme of work. Then there is the kind of evaluation which seeks to determine whether a student has acquired a particular skill or level of competence. Examples of this are a practical music examination, a swimming test, or an assessment of a student's teaching practice. Another kind of assessment is for the purpose of providing feedback on progress, to the learner or to other interested people – parents, for example. This could well involve self-assessment, and could include written or oral reports. Large-scale surveys, such as the work of the Assessment of Performance Unit (e.g. 1982, 1985) are intended to provide information about a whole population rather than about individual performances. The results are of use in obtaining a global picture of competence in a curriculum area and may influence educational policy, as indeed has been the case with the GCSE.

The purpose of some assessments is such that comparisons have to be made between people and a hierarchy established, whereas in other cases this is not necessary. One might be making an assessment of swimming competence, for example, in order to select the competitors for an inter-school gala, and generally the *fastest* swimmers will be selected. On the other hand, the reason for the assessment might be to decide whether a particular pupil is a sufficiently good swimmer to be able to take up dinghy sailing. In the latter case there is no need for comparisons or a hierarchy – unless there are more keen dinghy sailors than can be accommodated at the sailing school! In this situation the decision might be to give preference to the *strongest* swimmers. Changes in the purpose of assessment, then, can affect not only the need for selection but also the basis of selection.

Teachers have traditionally been involved in many of these kinds of assessment – e.g. writing end-of-term reports for parents, conducting school tests and examinations, discussing children's progress with parents or the pupils themselves, evaluating and comparing pupils in order to decide on setting arrangements, and so on. Others, such as national surveys, are more the province of researchers.

The general trend towards student records and profiling is in part owing to the wish to make fuller use of the kind of informal assessment which is in any case taking place.

Formal assessment of individual pupils for public examination purposes, however, has generally been conducted by the examination boards, although teachers have increasingly been involved in this. Somehow, many teachers are perfectly happy to carry out all sorts of informal assessment, but are worried by the notion of having to assess pupils' mathematics coursework for this more formal purpose, although, in fact, this is no more than an extension of a familiar task. This point needs to be emphasised, and teachers reassured that new skills are not required for the assessment of coursework. Rather, teachers will be required to utilise existing skills for a new purpose.

This chapter brings together and summarises the available information on the assessment of GCSE mathematics, with particular emphasis on coursework and investigative work. There are several illustrative examples of assessment of investigative work. Much of the rest of the information is purely descriptive – what the requirements are, what the terms mean. Also included is a description of GAIM, a graded assessment scheme which might be of interest to GCSE teachers.

THE ASSESSMENT REQUIREMENTS FOR GCSE MATHEMATICS

The assessment requirements for GCSE mathematics can be quickly summarised. These are laid down in section 6 of the *National Criteria – Mathematics*. They specify:

1. that there must be differentiated elements to meet the needs of different groups of candidates;
2. that at least one end-of-course written examination must be included in each scheme, and should normally account for at least 50 per cent of the assessment;
3. that for 1988, 1989 and 1990 an optional coursework element must be offered by each examining group;
4. that from 1991 onwards all schemes of assessment must include a coursework element; and
5. that where a coursework element is offered this must account for at least 20 per cent of the assessment (and by implication, normally not more than 50 per cent).

Chapter 2 of this book includes a summary of the examining groups' schemes of assessment, and it can be seen that the proportion of the assessment accounted for by coursework ranges from 25 per cent (which is little more than the minimum laid down) to 50 per cent, which is the recommended maximum.

Coursework, says paragraph 6.5 of the *National Criteria – Mathematics*, 'may take a variety of forms including practical and investigational work; tasks should be appropriate to candidates' individual levels of ability'. 'A variety of forms' is apparent in the differing coursework requirements of the examining groups which are also summarised in Chapter 2.

Terminology

Differentiated assessment

Differentiated assessment simply means:

1. that examination papers of different levels of difficulty and drawing on syllabuses with different amounts of curriculum content will be offered, and

2. that coursework tasks, where these are included, should be appropriate to candidates' individual levels of ability.

The other terms relating to assessment which you are likely to come across are: limited-grade examinations, grade descriptions, and grade-related criteria.

Limited-grade examinations

It is not difficult to understand what is meant by limited-grade examinations. Quite simply, it is that at each level of assessment there will be a limit, from above, from below, or from both, to the grades which may be awarded. The table in Chapter 2 (Figure 4) shows SEG's arrangements for grades as related to levels of assessment. It follows that any SEG candidate will necessarily be entered for a limited-grade examination, because the full range of grades is not available at any of the levels. A candidate at Level 2, for example, can only be awarded grades C, D, E or F. Grades A, B and G cannot be achieved, that is, a candidate who does not do well enough to achieve grade F will fail, and even an outstanding performance cannot result in the award of grades A or B. All of the examining group except NEA have proposed (and have had accepted) limited-grade examinations with ranges of grades similar to that of SEG. NEA's proposals, however, are for examinations limited from above at the lower and middle levels, and a full range of grades at Level R, the highest level. At the time of writing, this had not been approved by the Secondary Examinations Council.

The need for limited-grade examinations arises out of that central principle of GCSE, that candidates must be enabled to demonstrate what they know, etc. An examination which allows for the award of the full range of grades cannot do this. Weaker candidates would find little within their grasp (as is currently the case with full-range CSE examinations) and more able candidates might find the examination unchallenging.

Limited-grade examinations require, as is discussed in Chapter 2, that teachers enter candidates at an appropriate level. Otherwise, pupils may fail to achieve the grades of which they are capable. Limited-grade examinations also make it possible – as is indeed the intention within the GCSE scheme – to provide papers on which candidates can be required to achieve high marks. Thus, to reiterate one of the features of the new system which is spelled out in Chapter 2, the average candidate (i.e. the candidate who achieves the target grade at a given level) will be expected to get about two-thirds of the marks available. The grade below will be awarded for about half, and the grade above for about three-quarters of the marks available.

Grade descriptions

Grade descriptions are provided in the *National Criteria – Mathematics* and are intended 'to give a general indication of the standards of achievement likely to have been shown by candidates awarded particular grades'. In fact, grade descriptions are provided for grades C and F only, and we are expected to interpolate and extrapolate from these. For example, we are told that.

> Grade F candidates are likely to have shown a good knowledge of the subject content contained in List 1 and would be familiar with the associated processes and skills. They are likely to be able to apply this knowledge to single-concept problems of a type previously encountered.

It would follow from this that grade G candidates would show an adequate, but more patchy knowledge of the List 1 content etc, and that grade E candidates would show an excellent knowledge of this and the associated processes etc.

Appendix B contains extracts from this section of the *National Criteria – Mathematics,* including the specific examples on certain topics taken from the common-core lists, which are also provided.

A working knowledge of the different levels and grades can probably be acquired most easily by working through the specimen papers and then considering the available grades and the proportion of the papers which would need to be answered correctly to obtain a particular grade. For example, Papers 2 and 3 in the SEG scheme are targeted at grade D, which is to be awarded to a candidate who obtains about 67 per cent, while a grade C would be awarded for about 75 per cent and a grade E for about 50 per cent.

As grades will be awarded for overall performance, areas of relative strength and weakness may balance out. This means that it is very difficult, from one topic only, to assess a candidate's likely grade – one has to look at each candidate's attainment as a whole.

Grade-related criteria

For greater detail on what different (or at least key) grades mean in terms of achievement, we need grade-related criteria. This is perhaps the most difficult to understand of these technical terms, not least because grade-related criteria are still in the development stage. Another difficulty is that other technical terms – 'domains', 'criteria referencing' and 'norm referencing' in particular – tend to be used in discussions related to this.

In Chapter 1 there is a full discussion of the terms 'criteria referencing' and 'norm referencing'. One important aspect of the development of the GCSE examinations is, as is explained there, the shift from a primarily norm-referenced to a primarily criteria-referenced grading system. To some extent this arises out of a partial change in the underlying purpose of public examining at 16+. The main function originally was the establishment of a hierarchy of excellence on the basis of which important decisions could be made in a competitive market. Nowadays a function of equal weight is that users of the system (pupils, employers, parents, further and higher education) should be provided with accurate information on the standards which candidates have achieved.

The development of grade-related criteria is necessary for this latter function. In a criteria-referenced test, such as the driving test or a test of swimming competence, there must be objective (or as near as possible to objective) standards which a person must reach if he or she is to pass the test. These can be couched in terms such as 'can execute a three-point turn', 'can start on a hill', 'can swim twenty lengths' etc. These are the *criteria* by means of which it is decided whether someone has passed. In an examination such as GCSE, which has a range of pass grades, the statements of achievement (in other words, the criteria of success) must be related to the grades which can be achieved, hence the term 'grade-related criteria'.

The development of these criteria for mathematics is a major, and by no means simple, task. In July 1984 working parties in ten subjects were formed by invitation from the Secondary Examinations Council and were asked 'to establish criteria for the award of grades F, C and A across the domains'. A 'domain' was defined as 'a collection of the elements of a subject that form some reasonably coherent subset of the skills and competences needed in the subject' and each working party was asked to identify up to six 'domains'.

In September 1985, when the mathematics Working Party produced draft grade criteria, it recommended two domains. Domain 1 is called *Concepts, Skills and Procedures* and Domain 2 is called *Strategy, Implementation and Communication*.

Domain 1 'covers mathematical knowledge, skills and procedures together with their use to solve problems of standard types' while Domain 2 is 'concerned with the ability to think mathematically and with the use of mathematical strategies and processes in situations in which the approach to be used is not immediately apparent'. However, the Working Party is careful to point out that 'for purposes of assessment, artificial distinctions should (not) be made between them', adding that:

> although some assessment tasks will contribute entirely to assessment in one or other of the two domains, other tasks will be able to contribute to assessment in both domains.

It also comments that:

> the boundaries between domains may vary according to grade level. This is because problems and other mathematical tasks which, for those who achieve grade F, would fall within Strategy, Implementation and Communication may well be routine, and so contribute to assessment of Concepts, Skills and Procedures, for those who will achieve higher grades.

A similar point to this is made elsewhere in this book, that is, that the 'same' task may provide varying challenges for pupils of different attainment, depending on how it is tackled and how far it is developed.

At the time of writing, it is expected that grade-related criteria will be incorporated into GCSE syllabuses no sooner than 1991 (the year in which coursework becomes compulsory) and probably not until a year or two later than that. At this stage, therefore, it is only necessary to be aware that they are being developed and will be made available in due course.

GRADED ASSESSMENT, PROFILING AND CERTIFICATES OF ACHIEVEMENT

A parallel development to that of grade-related criteria is the development of various graded assessment schemes. Some are particularly designed to meet the needs of profiling and student record keeping or are especially aimed at the lower attainment levels (i.e. the 40 per cent of the school leaving population for whom GCSE, at least initially, is not intended). Examples are: OCEA (Oxford Certificate of Educational Achievement); SSCC (SMP, Suffolk LEA, Chelsea College, COSSEC) Graduated Assessment in Mathematics Study; and GAIM (Graded Assessment in Mathematics). GAIM is described below.

The GAIM scheme

The GAIM scheme is of particular interest to GCSE mathematics teachers on two counts. First, because the work being done by the GAIM researchers is contributing to the development of grade-related criteria for GCSE and, second, because it is proposed that each of the seven later GAIM levels (there are fifteen in all) should correspond to one of the grades of GCSE. The intention is: 'to produce a mathematics assessment scheme appropriate for all 11–16-year-old students in comprehensive schools' and 'to provide a detailed cumulative record of a student's attainment in mathematics.

Assessment within the GAIM scheme has four inter-related components: extended pieces of work, practical problem-solving, investigations, and topic criteria. Investigations and practical problem-solving are intended to reflect the pure and applied aspects of mathematics respectively, and topic criteria are knowledge, understanding and skills expressed as 'can do' statements and classified by level of difficulty and by topic.

Materials for teaching and assessing these at each level are being developed and piloted in several dozen schools. The GAIM scheme is also using materials from published sources, and aiming to place these appropriately within the scheme.

Several of the pieces of work in Chapter 4 were carried out as part of GAIM piloting. The Winston's garage worksheet was produced by GAIM. Planning a kitchen is a SMILE resource which is being tested for GAIM, and Monica Ramsay carried out this work with her pupils in collaboration with the GAIM researchers.

GAIM is also producing a series of video units which 'sets out to provide a resource bank of observation material on various aspects of teaching and learning mathematics'. These are very relevant to the GCSE as they focus on, for example, investigations, practical problem-solving and collaborative work.

ASSESSING INVESTIGATIVE WORK

In Chapter 3 the two classroom packs produced by the Shell Centre for Mathematical Education and the Joint Matriculation Board are discussed. A valuable aspect of the packs is the guidance they give on assessing examination questions which are more open-ended than traditional questions. A number of specimen questions are provided in each of the packs, and several of these are reproduced in the *Guide for Teachers* produced by the Secondary Examinations Council in collaboration with the Open University (1986). A suggested mark scheme is provided for each of the specimen questions, together with sample pupil work and a discussion of the marking of the actual work produced.

Studying this sort of resource is very helpful, but the point which is made about 'secondhand' problem-solving in Chapter 3 – i.e. reading about problem-solving as opposed to engaging actively in this – is equally applicable here. There is no effective substitute for the activity of assessment – you simply have to try it out, preferably in collaboration with colleagues. Until you have had the experience of assessing coursework or investigative work for yourself, you will find it hard to believe that it is really perfectly straightforward and far less difficult than you might have believed.

The Matchstick Investigation

As a halfway stage, you might like to try an experiment based on the Matchstick Investigation. This investigation was part of the SMILE CSE examination this year (1986), and was then given to the fourth-year pupils at a girls' comprehensive school as part of their end-of-year examination (The question is reproduced as Figure 1.) Monica Ramsay's pupils carried out this investigation in a double period. As it was an examination, the girls were expected to work individually and in silence.

It is worth commenting here that the SMILE GCSE submission does not include investigational work carried out under examination conditions, except at the highest level. Investigations, carried out as part of pupils' ordinary classroom activities, will be a component of the coursework. This is because it is felt to be contradictory, in general, to expect open-ended work to be done in the closed situation of a timed examination. Also, experience has shown that setting an *examination* of this type has a detrimental effect on classroom practice, as teachers tend to train children for this. Nonetheless, the experience amassed by SMILE teachers over some years of formally assessing investigational work is of use to the rest of us. Whether the work was done under examination conditions or in ordinary lessons, the same criteria, more or less, would be used in marking it.

1 Matchsticks

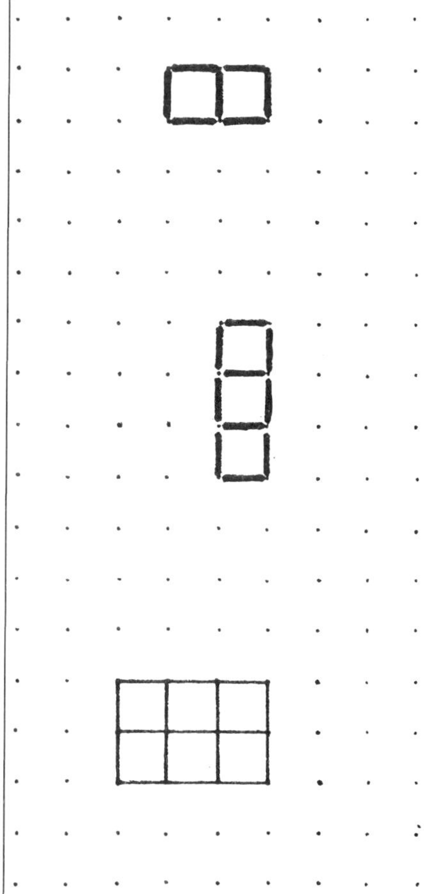

To make this grid you would need 7 matches.

To make this grid you would need 10 matches.

To make this grid you would need 17 matches.

Investigate for different size grids.

▲ *Fig. 1 Instruction from the EGA 4th Year Exam Mathematics 2 paper, Investigations*

How to proceed

It is recommended that you treat this as a group activity, to be shared with as many colleagues as you can persuade to participate. Before any of you read the suggested marking scheme, have a go at the investigation yourselves, working individually at this stage. Discuss your work, and agree on a mark scheme. Once this has been done, you might find it interesting to compare your mark scheme with the one the teachers at this school used (see below). In the light of this, would you want to amend yours? Or would you advise them to amend theirs?

There is no absolute right or wrong about a mark scheme like this. Quite often, one prepares a mark scheme which relates to the way it is expected that students will tackle a task and, in the event, some of them see things in a different way. The mark scheme might then need to be amended so as to give appropriate credit to their work. As is stated elsewhere, absolute objectivity is neither possible nor necessary.

The next stage would be to consider how to introduce the activity to pupils so as to have some work on which to try out your mark scheme.

Suggested marking scheme for the Matchstick Investigation

Results   3 marks –
1 mark for each correct example – these must be different from the examples on the sheet. Rotations do not count as different.

Order   1 mark – Any type of order or grouping of results, 1 mark.

Table   1 mark – Any table, 1 mark.
Both these marks can be for either a very good table (with results in order) or for results in an order and well presented.

Observation of pattern   maximum 2 marks
Any pattern, for example a mapping ($x \rightarrow x + 3$ for single rows) or a statement to that effect or any comment about odd and even patterns etc., 1 mark per comment, maximum 2.

Generalisation   Any number of marks to bring total to 10.
A specific comment in words or algebra, e.g. $M = 3s + 1$ for single rows. 1 mark (+1 pattern if they have not already got 2) or a complete generalisation (i.e. one which works for any grid).

Comments

As explained above, it would not be appropriate to give your pupils this (or any) investigation to do under examination conditions, especially if they are new to this kind of work. Not only does it take time for children to learn to work in an investigative way (the pupils whose work is reproduced had had quite a lot of experience of this), but also it is, arguably, inappropriate ever to expect investigative work to be done in a timed examination. However, now that your (and hopefully one or more colleagues) have worked with the matchstick investigation, you may wish to offer it to some of your classes. You could organise it as a collaborative activity, with pupils perhaps working in groups of two or three. Try to avoid telling the children what to do, but instead encourage them to ask their own questions, look for patterns, make notes as they work, organise their work clearly and perhaps draw up a table of results as an aid to pattern-spotting.

Generally speaking, it is better not to 'finish' an investigation before offering it to pupils, as this could be inhibiting. The hardest thing for a teacher is to keep quiet and let children work things out for themselves, and it is very tempting, if you think you know all the answers, to want to help your pupils find them! However, it is probably a good idea to 'have a go' first, and certainly to have taken the investigation quite far yourself before giving it to children when you are new to this way of working. It could be quite terrifying to give an investigation cold to children if neither you nor they have experience of working like this. Also, as investigations should be fun for everyone, do not offer any to your pupils which you find dull. So if the matchstick investigation does not interest you, choose something else.

The final stage in this experiment would be to get together with your colleagues and as a joint activity assess all the pupils' work and place it into an order of merit. According to most schemes, GCSE coursework assessment is to be done by teachers and externally moderated. The moderators will only see a proportion of the work, and teachers' primary responsibility will be to arrive at an order of merit for all pupils in a school. If your marking is more fierce than most, all your marks will be upgraded so that your pupils are not disadvantaged relative to candidates from other schools. On the other hand, if you are too lenient, all the marks will be lowered! In this way, comparability between schools can be maintained without the group's examiners having to mark every piece of work.

There are, however, many different patterns. SEG, for example, is planning, where possible, to organise the assessment of coursework through moderating committees composed of groups of local teachers. If you choose its syllabus, then, you may have the opportunity to participate in moderating the work of pupils from several schools. The remarks above remain valid, however, as moderators, whether they are local teachers or examiners appointed by the group, will see only a proportion of the work. The onus will still be on each school to arrive at a merit order for its own candidates. This is perfectly right and proper, as only teachers can be in a position to judge one candidate at a school against another. Teachers, and only teachers, will know how much help a candidate has had, or whether a pupil has displayed particular qualities of, e.g. individuality and imagination, relative to others in the school.

Thus, working with colleagues on techniques of assessment is of vital importance. It is not particularly important that the final assessments should be absolutely 'objective' – i.e. that you ensure that other people marking the same work with the same mark scheme would have given identical marks. What will matter in the assessment of coursework and other open-ended work is that you have learned to judge qualitative differences

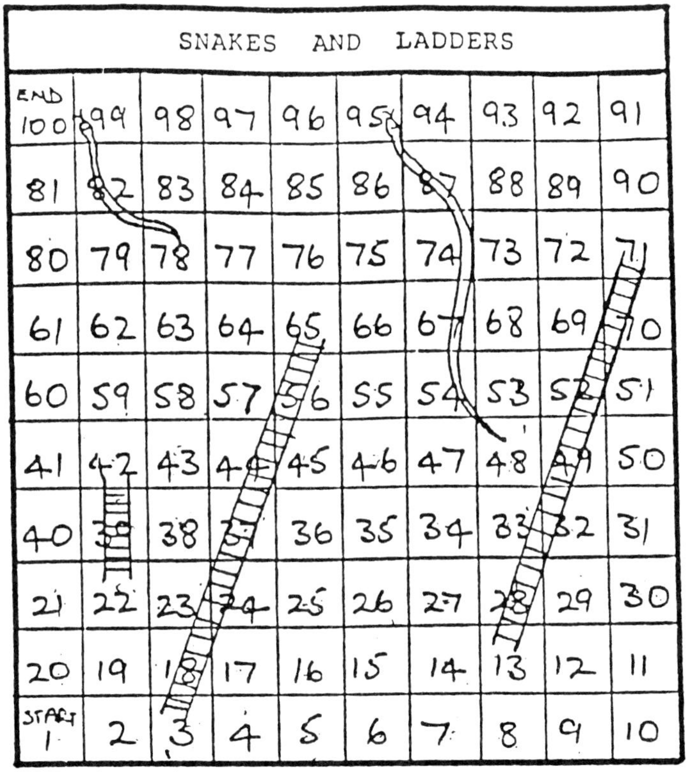

SNAKES AND LADDERS

In the game of Snakes and Ladders a player moves according to the throw of one dice. If she lands at the bottom of a ladder she climbs up to the top. If she lands at the top of a snake she slides down to the bottom.

Two of the ladders shown are of the vector form $\begin{pmatrix} 2 \\ 6 \end{pmatrix}$

A player landing on 3 would climb to 65.

So 3 ⟶ 65

and 13 ⟶ 71

Investigate where you climb to with different starting points.....different ladders.

▲ *Fig. 2*

between pieces of work and are able to arrive at a consensus within your school. For this, there is no substitute for practice. Some further suggestions for acquiring the necessary experience follow.

The Snakes and Ladders Investigation

This too was set in a SMILE examination paper (1986, O level), see Figure 2. Tony Purcell, one of the contributors in Chapter 4, then gave it to some of his fourth-year pupils. An example of the work produced is shown in Figure 3. The girls did the work under examination conditions with a 45-minute time limit. The exercise was used as part of the school's GCSE training, and Tony Purcell comments that it generated lively discussions – and disagreements – among pupils and teachers. It was questioned whether it was fair to place a time restriction on an investigation of this nature and therefore whether this is an appropriate piece of work for an examination. These are very apt criticisms – readers are referred to the discussion above.

Once again, you might like to make use of this, or in fact any investigation of your choice, to try out this style of working and then to practise assessment. Many of the projects and activities which are described in Chapter 4 lend themselves to this – Martin Marsh comments in his account that he found the Chinese triangles project very useful for practising his own assessment skills, and this work, of course, was carried out over a period of time, in class and as homework.

▼ *Fig. 3*

```
Kelly O'Keeffe. 4H
                                July 8th. 1986.      20 → 64      40 → 84      60 → ...
                                                     19 → 65      41 → 85      ...
Using the vector (6/2):                              18 → 66      42 → 86      ...
       s    f            s    f           s    f    17 → 67      43 → 87      ...
  a. 1 → 27       b. 21 → 47       c. 41 → 67       The two rules do not work if the
     2 → 28          22 → 48          42 → 68       numbers in the vector are odd.
     3 → 29          23 → 49          43 → 69
     4 → 30          24 → 50          44 → 70       Using the vector (3/6):

     odd rows = 10y + x + 5 = f.                     1 → 64
                                                     2 → 65         10y + x + 5 = f
                                                     3 → 66
    20 → 34      40 → 54      60 → 74               4 → 67
    19 → 33      39 → 53      59 → 73
    18 → 32      38 → 52      58 → 72               20 → 77
    17 → 31      37 → 51      57 → 71               19 → 76        10y - x + 5 = f
                                                    18 → 75
     even rows = 10y - x + 5 = f.                   17 → 74

Using the vector (3/5):
                                                    But if the y number in the vector
    1 → 57      21 → 77      41 → 97                is even, the rules will work.
    2 → 56      22 → 76      42 → 96                ∴ If you always keep the y number
    3 → 55      23 → 75      43 → 95                even, the rules will always work, and
    4 → 54      24 → 74      44 → 94.               it doesn't matter if the x number is odd
                                                    or even.
```

A very useful resource for teachers developing their assessment skills is the set of booklets produced by the West Sussex Institute of Higher Education and the Southern Regional Examination Board. TEAM (Teachers Evaluating and Assessing Mathematics) consists of six sections along with an introductory booklet. They are: *Pupils' Work, Reflection, Changing the Climate, Getting Started, Evaluation* and *Assessment.* Although not produced in response to the new examination, this material is of particular relevance to GCSE teachers getting to grips with the new assessment demands.

ASSESSING COLLABORATIVE WORK

Probably the most difficult kind of work to assess with a reasonable degree of accuracy is an individual's contribution to a collabortive project. If this is carried out as ordinary classwork, as in the case of the statistics project which Bridget Perkins describes in Chapter 4, a rough assessment based on your knowledge of what the children have been doing is probably adequate. If a formal assessment is needed, you will need to talk individually to each child, in other words carry out an oral assessment, and also probably require individual write-ups of the work. This leads into the final section of this chapter.

ORAL ASSESSMENT AND AURAL TESTS

Most of the examining groups specify that oral assessment is to form part of coursework assessment, and some provide for aural tests as well – details are given in Chapter 2.

Aural tests

Aural assessment is perfectly straightforward. The teacher reads the questions which have been set by the examining group, and the pupils write down their responses. The questions may involve mental calculations, reading information from tables or diagrams provided, etc. They are different from standard written papers only in that candidates have to *listen* to the questions instead of reading them. Also, the short time allowed is intended to encourage any working to be done 'in the head' – although generally doing some pencil-and-paper working-out is not forbidden, just discouraged. Marking these aural tests is no different from marking any written test.

Oral assessment

In contrast to this, oral assessment is an entirely verbal interaction. The teacher might ask a pupil to explain something in a piece of work he or she has done, and assess how clearly and accurately the explanation is given. Or the teacher may 'listen in' to a discussion between two or three pupils, and assess the contribution each is making. There is possibly less clarity about how the examining groups expect oral assessment to be carried out than any other element. My view is that teachers are, or ought to be, making oral assessments of their pupils all the time in the ordinary course of classroom work.

Chapter 3 draws attention to the critical importance of oral work in the development of mathematical concepts and knowledge. A great deal of this work will necessarily be carried out without the teacher's active participation – e.g. when the class is divided into small groups. From time to time, however, it would be natural and appropriate to join a group of pupils, listen to their discussion, perhaps ask questions designed to take them on in their explorations. It is at these times, and also when the teacher is talking with individuals about their work, that informal assessments can be made of a child's oral competence in mathematics. Quite often, a child will surprise the listener by verbalising in an unexpectedly sophisticated way. On the other hand, what a child has said may point to a particular misconception which had previously gone unnoted.

Keeping a record

Whatever the nature of the interaction, it would be helpful, in terms of building up a profile for that child as well as for future formal assessment purposes, to make a note of the conversation as soon as possible. Perhaps a small notebook, permanently available, would make it possible to jot down a quick reminder which could be written up more fully on the child's record sheet later. The crucial point is that the oral assessment should not be regarded as an additional task over and above the job of teaching. Keeping a formal record may be a new task, but actually doing the assessment in the classroom should happen whether or not records are required.

Undoubtedly, keeping such a record would be extremely valuable – even if it is a chore that busy teachers could do without. Some people have superb memories and can keep the detailed information they have built up on all their pupils 'in their heads'. Most of us are not so fortunate and need a little help. Also, a record like this would be an extremely useful addition to the usual mark sheets when a new teacher takes over a class.

Quite apart from the formal requirements in the GCSE, records of oral assessment could appropriately be regarded as an aspect of the normal out-of-classroom duties of any teacher, with time being allowed for keeping them up to date.

SUMMARY

Different kinds of mathematical activity and classroom organisation together with new assessment requirements add up to the need for teachers to have time to work collaboratively with colleagues and to keep full and varied records in addition to the normal duties of lesson preparation and marking.

There is also a pressing need to communicate with parents and help them to understand the new ways of learning mathematics which their children are experiencing. This too requires time and thought. In Appendix A, Deanne Reynoldson describes a parents' workshop which she organised for this purpose.

The picture which is built up through considering the implications of the new examination system is one where mathematics teaching is much more varied and exciting than has traditionally been the case. If teachers are to do justice to the demands placed on them, they will need more time outside the classroom than is generally allowed at present. There is much to be learned, much to be practised and much fun to be had along the way.

Appendix A

Deanne Reynoldson

GCSE MATHEMATICS – A PARENTS' WORKSHOP

Our new school, a girls' comprehensive in inner London, was formed two years ago through an amalgamation of two schools. It seemed an ideal opportunity for doing something innovative, so we decided to pilot the OMEGA scheme. (Author's note – see description of OMEGA in Chapter 4). At this time, teachers' feelings ranged from excitement and enthusiasm to open scepticism.

Once the initiation of the teachers and pupils was under way, the next step was to provide an opportunity to discuss with parents the mathematical education their daughters were experiencing. As a result of industrial action, however, the schemes had been operating for nearly two years by the time the first opportunity to do this occurred. During this period parents had questioned our new way of working and asked why we were not working in ways with which they were familiar, but our pupils' enthusiasm for their mathematics lessons had given us the confidence to continue.

As soon as the industrial action was suspended, my concern was how to organise the meeting with parents effectively. I considered the following; a talk on the aims of the department and the change of emphasis in mathematics teaching since the Cockcroft Report; showing parents a video on investigational work; inviting parents to visit pupils while they were working in their classrooms.

Gradually my ideas crystallised, and what emerged was that the way to initiate parents was by giving them experience of the essence of what we were trying to do with the children. This would be by getting them to:

– experience collaborative learning by working in groups;
– carry out practical activities;
– investigate;
– discuss what they were doing, justify it, and as a result experience talking about mathematics.

The decision was therefore that the afternoon's programme would consist of two activities; a practical problem-solving activity and an investigation. Parents would work with groups of first- and second-year pupils, including their own daughters. We were pleased that 35 parents decided to attend the workshop, and decided to have first-year pupils join in for the first activity and second year pupils for the second.

Activity 1

For the first activity, the introductory lesson to the module on volume seemed ideal. The problem can be very simply stated: Using a piece of thin card of A4 size to form the sides of a container, find the maximum amount of macaroni it will hold.

The materials provided for each group were: A4 pieces of card, approximately $2\frac{1}{2}$ kg of macaroni, a container for the macaroni, a scoop for measuring, Sellotape, and pairs of scissors.

Once each group of parents and first-year pupils was settled with their materials and a card stating the problem, they were left to work on their own. The teachers present moved from table to table, countering question with question. It was interesting to note the confidence of some of the adults in trying to cut short the activity for the girls.

For example, one parent was overheard to tell his daughter when she suggested changing the container to give it a circular instead of a rectangular cross-section 'Don't be silly! Of course it will hold the same amount.' Fortunately, most of the groups were involved in disproving this assertion, and many found that a cylinder formed by joining the shorter sides of the card held more than if the longer sides were joined. One group even cut the card into strips, while another insisted on making envelope shapes. All the groups devised a standard measure, some of them discussing this heatedly. Two girls worked out the average number of pieces of macaroni in a scoopful by finding the mean for five scoopfuls. Their results were recorded in pieces of macaroni.

Discussion about the relationship between this problem and the packaging industry concluded the activity.

Activity 2

After tea, the second-year pupils joined the parents. The second activity was begun after an introductory discussion about the purpose of investigations. Each group was given a sheet which explained the task. (See Figure 1.)

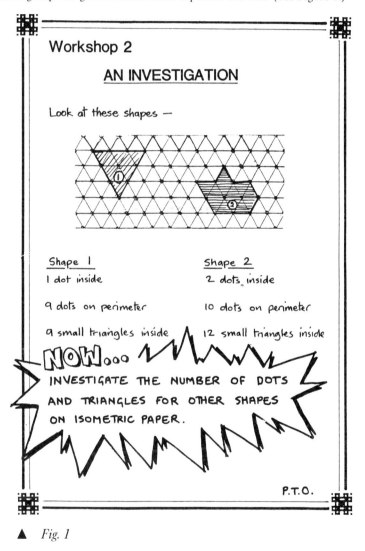

▲ *Fig. 1*

This proved to be a much more difficult activity. Most of the parents were able to do many examples, and some set out tables of results. We allowed about 40 minutes for the task, and many of the pupils wanted to continue this investigation for homework. Five of the 39 girls in the group found a complete generalisation, and so did one parent. On the whole, however, the results showed that pupils needed more practice in organising results systematically and looking for patterns.

Conclusion

The last part of the afternoon was given over to questions and answers. Sometimes pupils wished to respond to the questions posed by parents, and it was rewarding to note their confidence in and support for the work they had been doing.

In all, we felt that the afternoon had been a great success. Parents were clearly very interested in trying to understand the changes that are occurring in the teaching of mathematics, and requested further workshops. This is especially the case now that they understand that these changes meet the requirements of the GCSE examination, where discussion, practical work, problem-solving and investigational work have become an integral part of the course.

Had we run the workshop in the evening, it is possible that more parents would have been able to attend. We felt, however, that it was an invaluable experience to have our pupils work with the visitors during normal school hours, and we would like to repeat this. In an effort to reach more parents, we wrote an account of the workshop for our termly parents' magazine and included some of the questions (and answers) which we considered would be of interest to most parents. Some of these are set out below.

Q. Do you think it is right that pupils should mark their own work?
A. For the type of question generally asked in mathematics, yes. The answers are usually right or wrong, so the pupils quickly find out whether they are tackling the task correctly, and can seek help if necessary.
Q. What if they just copy the answers? Can you trust them to mark their own work?
A. Part of the process of being educated is learning to be responsible. To use the answer book as an aid to learning is a mature, responsible act. The pupils know if they are cheating. They do not feel good about it and they are only cheating themselves.
Q. Does the sort of practical work we have been doing today lead to your giving them the standard formula so that they can get on with doing examples in the traditional way?
A. No, it does not. The underlying aim of this type of work is that pupils come to devise their own rules, that is, something which will work for them so that they can solve the problem. As they become more experienced, their formulae become more sophisticated. It is at this point that teachers use their expertise to judge whether a pupil can cope with abstract formulae. The important point is that the pupils themselves have found something which works for them. They have not been given a meaningless result and then told to use it.
Q. What use is all this sort of maths when you leave school?
A. I would say that there are least three uses.

  First, we are aiming to develop an understanding of mathematics as something which is inextricably involved in so much of what we see and do. Mathematics is all around us, so we must provide opportunities for our pupils to come to realise this.

  Second, it is important for our pupils not only to gain numerical, spatial and estimation skills but also to have developed strategies for problem-solving as a vital aspect of their everyday lives.

  Third, investigations provide an opportunity for pupils to gain insight into what it is like to be a mathematician. We want this experience to be available for many more pupils than in the past, so that they can 'get on the inside' of mathematics instead of operating by using rules that they do not understand. Even when they do understand traditional methods, they should have sufficient confidence in themselves as mathematicians to believe that their own methods can be equally successful. Schools cannot teach *all* the skills for work. Each job requires different strengths, and when the confidence and motivation are there, an employee can learn very quickly. What we can do is try to provide an opportunity to develop an understanding and appreciation of the function and fascination of mathematics, together with an interest in the subject that stays with them beyond the classroom door.

The parents' questions highlighted many important issues. We believe that a forum in which these can be discussed is essential if we are to develop the right support for the children who are our common concern.

Appendix B
Extracts from *The National Criteria – Mathematics*

4. Content 4.1 In any differentiated scheme offering the full range of grades there will normally be at least three levels of assessment. For each of these levels a separate content must be prescribed. This will consist of 'core' items, common to the syllabuses of all Examining Groups, together with additional material chosen at the discretion of the Group devising the syllabus. The 'core' items are set out in **List 1** and **List 2**.

4.2 The content of List 1 must be included in all syllabuses which bear the title Mathematics. It should constitute almost the whole of the syllabus content for any examination on the results of which the great majority of candidates is expected to be awarded grades E, F or G.

4.3 The content of both List 1 and List 2 must be included in the syllabus for any examination which can lead to the award of grade C. These lists should constitute almost the whole of the syllabus content for any examination on the results of which the great majority of candidates is expected to be awarded grades C, D or E.

4.4 Pupils aspiring to the award of grades A or B will be expected to have covered a syllabus well in excess of that contained in List 1 and List 2. When choosing the additional material to be included, Examining Groups should have regard to the need both to enable pupils to develop their powers of abstraction, generalisation and proof and also to provide a firm foundation for future work in mathematics at A-level and beyond. In any scheme of assessment at this level, between 50 per cent and 70 per cent of the marks allocated to written examination papers should be carried by questions which relate to items in List 1 and List 2.

List 1	**List 2**
Whole numbers: odd, even, prime, square. Factors, multiples, idea of square root.	Natural numbers, integers, rational and irrational numbers. Square roots.
Directed numbers in practical situations. Vulgar and decimal fractions and percentages; equivalences between these forms in simple cases; conversion from vulgar to decimal fractions with the help of a calculator.	Common factors, common multiples. Conversion between vulgar and decimal fractions and percentages. Standard form.
Estimation.	Approximation to a given number of significant figures or decimal places.
Approximation to obtain reasonable answers.	Appropriate limits of accuracy.

List 1 *continued*	**List 2** *continued*
The four rules applied to whole numbers and decimal fractions.	The four rules applied to vulgar (and mixed) fractions.
Language and notation of simple vulgar fractions in appropriate contexts, including addition and subtraction of vulgar (and mixed) fractions with simple denominators.	
Elementary ideas and notation of ratio. Percentage of a sum of money. Scales, including map scales. Elementary ideas and applications of direct and inverse proportion. Common measures of rate.	Expression of one quantity as percentage of another. Percentage change. Proportional division.
Efficient use of an electronic calculator; application of appropriate checks of accuracy.	
Measures of weight, length, area, volume and capacity in current units. Time: 24 hour and 12 hour clock. Money, including the use of foreign currencies.	
Personal and household finance, including hire purchase, interest, taxation, discount, loans, wages and salaries. Profit and loss, VAT. Reading of clocks and dials. Use of tables and charts. Mathematical language used in the media. Simple change of units including foreign currency. Average speed.	
Cartesian coordinates. Interpretation and use of graphs in practical situations including travel graphs and conversion graphs. Drawing graphs from given data.	Constructing tables of values for given functions which include expressions of the form: $ax + b$, ax^2, a/x ($x \neq 0$) where a and b are integral constants. Drawing and interpretation of related graphs; idea of gradient.
The use of letters for generalised numbers. Substitution of numbers for words and letters in formulae.	Transformation of simple formulae.
	Basic arithmetic processes expressed algebraically. Directed numbers. Use of brackets and extraction of common factors. Positive and negative integral indices.
	Simple linear equations in one unknown. Congruence.
The geometrical terms: point, line, parallel, bearing, right angle, acute and obtuse angles, perpendicular, similarity.	
Measurement of lines and angles. Angles at a point. Enlargement.	Angles formed within parallel lines.

List 1 *continued*

Vocabulary of triangles, quadrilaterals and circles; properties of these figures directly related to their symmetries. Angle properties of triangles and quadrilaterals.

Simple solid figures.

Use of drawing instruments. Reading and making of scale drawings. Perimeter and area of rectangle and triangle. Circumference of circle. Volume of cuboid.

Collection, classification and tabulation of statistical data. Reading, interpreting and drawing simple inferences from tables and statistical diagrams. Construction of bar charts and pictograms. Measures of average and the purposes for which they are used.

Probability involving only one event.

List 2 *continued*

Properties of polygons directly related to their symmetries. Angle in a semi-circle; angle between tangent and radius of a circle. Angle properties of regular polygons.

Practical applications based on simple locus properties.

Area of parallelogram. Area of circle. Volume of a cylinder.

Results of Pythagoras. Sine, cosine and tangent for acute angles. Application of these to calculation of a side or an angle of a right-angled triangle.

Histogram with equal intervals. Construction and use of pie-charts. Construction and use of simple frequency distributions.

Simple combined probabilities.

7. Grade Descriptions

Grade descriptions are provided to give a general indication of the standards of achievement likely to have been shown by candidates awarded particular grades. The grade awarded will depend in practice upon the extent to which the candidate has met the assessment objectives over-all and it might conceal weakness in one aspect of the examination which is balanced by above average performance in some other.

Grade F candidates are likely to have shown a good knowledge of the subject content contained in List 1 and would be familiar with the associated processes and skills. They are likely to be able to apply this knowledge to single-concept problems of a type previously encountered.

Grade C candidates are likely to have shown a good knowledge of the subject content contained in both Lists 1 and 2 and would be able to execute accurately the processes and skills associated with the content of List 1. They are likely to be able to apply the knowledge, processes and skills to structured situations and show an ability to select a correct strategy to solve a multi-concept problem.

There will be no system of hurdles for a particular grade. Candidates will gain credit from any positive achievement in different sections of the syllabus.

Standards of achievement at Grades F and C in those assessment objectives which refer to knowledge, skills, applications and problem-solving (ie assessment objectives 3.1, 3.2, 3.12, 3.13, 3.14, 3.15 and when applicable 3.16) are contained in the general descriptions given above. Standards of achievement at Grades F and C for the other assessment objectives are shown below by *examples* on certain topics taken from the common core lists. A substantial majority of candidates in each of these grades will have demonstrated competence at the appropriate level in the examples given.

Grade Descriptions

Assessment Objective	Grade F examples	Grade C examples
3.3	Extract information from simple timetables. Tabulate numerical data to find the frequency of given scores. Draw a bar chart. Plot given points. Read a travel graph.	Construct a pie chart from simple data. Plot the graph of a linear function.
3.4	Add, subtract and multiply integers. Add and subtract money and simple fractions without a calculator. Calculate a simple percentage of a given sum of money.	Apply the four rules of number to integers and vulgar and decimal fractions without a calculator. Calculate percentage change.
3.5	Perform the four rules on positive integers and decimal fractions (one operation only). Convert a fraction to a decimal.	Perform calculations involving several operations, including negative numbers.
3.6	Measure length, weight and capacity using metric units. Understand relationships between mm, cm, m, km; g, kg.	Use area and volume units.
3.7	Perform a money calculation with a calculator and express the answer to the nearest penny.	Give a reasonable approximation to a calculator calculation involving the four rules.
3.8	Draw a triangle given three sides. Measure a given angle.	Use a scale drawing to solve a two-dimensional problem.
3.9	Continue a straightforward pattern or number sequence.	Recognise, and in simple cases formulate, rules for generating a pattern or sequence.
3.10	Use simple formulae, eg. gross wage = wage per hour × number of hours worked, and use of $A = l \times b$ to find the area of a rectangle.	Solve simple linear equations. Transform simple formulae. Substitute numbers in a formula and evaluate the remaining term.
3.11	Recognise and name simple plane figures and common solid shapes. Find the perimeter and area of a rectangle. Find the volume of a cuboid.	Calculate the length of the third side of a right-angled triangle. Find the angle in a right-angled triangle, given two sides.
	In schemes of assessment where the objective is applicable:	
3.17	Carry out a simple survey; obtain straightforward results from the information obtained.	Investigate and describe the relationship between the surface area and volume of a selection of solid shapes.

Selected Bibliography

APU (1982) *Mathematical Development*, Secondary Survey Report No. 3, London; HMSO.

APU (1985) *A Review of Monitoring in Mathematics, 1978 to 1982*, London: HMSO.

Banwell C S, Saunders K D and Tahta D G (1972) *Starting Points*, Oxford University Press. Republished and updated 1986, Tarquin (see address below).

Bestley, Frances (1986) 'No Circles' *Splash* No. 111110 (June). (Available from SMILE centre, address below.)

Bird, Marion (1983) *Generating Mathematical Activity in the Classroom*, West Sussex Institute of Higher Education. (Available from The Mathematical Association, see address below.)

Bolt, Brian (1982) *Mathematical Activities*, Cambridge University Press.

Bolt, Brian (1985) *More Mathematical Activities*, Cambridge University Press.

Burton, Leone (1984) *Thinking Things Through*, Basil Blackwell.

Burton, Leone (ed) (1986) *Girls Into Maths Can Go*, Holt, Rinehart & Winston.

Buxton, Laurie (1981) *Do You Panic About Maths?*, Heinemann Educational Books.

Castle F (1899) *Elementary Practical Mathematics*, Macmillan.

DES (1975) *A Language for Life*, London: HMSO (The Bullock report)

DES (1982) Committee of Inquiry into the Teaching of Mathematics, *Mathematics Counts*, London: HMSO (the Cockcroft Report).

DES (1985a) *General Certificate of Secondary Education: A General Introduction*, London: HMSO.

DES (1985b) *General Certificate of Secondary Education: The National Criteria – General Criteria*, London: HMSO.

DES (1985c) *General Certificate of Secondary Education: The National Criteria – Mathematics*, London: HMSO.

DES (1985d) *Mathematics from 5 to 16*, London: HMSO.

Flanders N A (1970) *Analyzing Teaching Behavior*, Reading, Mass: Addison-Wesley.

Floyd Ann (ed) (1981) *Developing Mathematical Thinking*, Addison-Wesley in association with the Open University Press.

Gardner Martin (1965) *Mathematical Puzzles and Diversions*, Pelican (first published 1959).

Godfrey C and Siddons A W (1946) *The Teaching of Elementary Mathematics*, Cambridge University Press.

Hart K (ed) (1981) *Children's Understanding of Mathematics: 11–16*, John Murray.

Harvey R, Kerslake D, Shuard H & Torbe M *Language Teaching and Learning: 6, Mathematics*, Ward Lock Educational.

Hoyles Celia (1982) 'The pupil's view of mathematics learning' in *Educational Studies in Mathematics*, Vol. 13, pp 349–72.

Kasner E and Newman J (1968) *Mathematics and the Imagination*, Pelican (first published 1940).

Köhler W (1925) *The Mentality of Apes*, New York: Harcourt.

Langdon N and Snape C (1984) *A Way with Maths*, Cambridge University Press.

Lakatos Imre (1976) *Proofs and Refutations: The Logic of Mathematical Discovery*, Cambridge University Press.

Mason John with Burton L and Stacey K (1982) *Thinking Mathematically*, Addison-Wesley.

Modular Mathematics Organisation (1986) *Maths in Context*, Heinemann Educational Books.

Mottershead, Lorraine (1978) *Sources of Mathematical Discovery*, Basil Blackwell (first published 1977).

Mottershead, Lorraine (1985) *Investigations in Mathematics*, Basil Blackwell.

Papert S (1980) *Mindstorms*, Harvester Press.

Pirie, Susan (1986) *Investigations in Your Classroom* (Available from Dr Susan Pirie, Department of Science Education, University of Warwick, Coventry CV4 7AL.)

Polya G (1957) *How To Solve It*, Anchor Books (first published 1945, Princeton University Press).

Practical Mathematics in Schools Project Group (1984) *Practical Mathematics in Schools*, Scottish Curriculum Development Service, College of Education, Gardyne Road, Broughton Ferry, Dundee DD5 1NY.

The Royal Society (1986) *Girls and Mathematics*, a report by the joint Mathematical Education Committee of the Royal Society and the Institute of Mathematics and its Applications. Available from the Royal Society, 6 Carlton House Terrace, London SW1Y 5AG.

Secondary Examinations Council (1985a) *Working Paper 1: Differentiated Assessment in GCSE*, available from the SEC, address below.

Secondary Examinations Council (1985b) *Working Paper 2: Coursework Assessment in GCSE*, available from the SEC, address below.

Secondary Examinations Council (1985c) Report of Working Party, *Mathematics: Draft Grade Criteria*, SEC.

Secondary Examinations Council (1986) *Working Paper 3: Policy and Practice in School-based Assessment*, available from the SEC, address below.

Secondary Examinations Council in collaboration with the Open University (1986) *GCSE Mathematics: A Guide for Teachers*, Open University Press.

Shell Centre for Mathematical Education/Joint Matriculation Board (1984) *Problems with Patterns and Numbers: an O-level module*, Manchester, Joint Matriculation Board. Available from the Shell Centre, address below.

Shell Centre for Mathematical Education/Joint Matriculation Board (1985) *The Language of Functions and Graphs: an examination module for secondary schools*, Manchester, Joint Matriculation Board. Available from the Shell Centre, address below.

Skemp, Richard (1971) *The Psychology of Learning Mathematics*, Penguin Books.

Skemp, Richard (1976) 'Relational understanding and instrumental understanding' in *Mathematics Teaching*, No. 77.

Spode Group (1983) *Solving Real Problems with Mathematics* (two volumes), Cranfield Press, Cranfield Institute of Technology, Bedford MK43 0AL.

University of London School Examinations Board (1985) *Sexism, Discrimination and Gender Biases in GCE Examinations*, University of London School Examinations Board.

Waddingham J and Wigley A (1986) *Secondary Mathematics with Micros Inservice Pack*, MEP. (Available from AUCBE Endymion Road, Hatfield, Herts.

Wells, David (1985) 'Problems, Investigations and Confusion' in *Mathematics in school*, Vol. 14 No. 1.

West Sussex Institute of Higher Education, *Teachers Evaluating and Assessing Mathematics* (TEAM), produced in conjunction with the Southern Regional Examination Board and available from the West Sussex Institute, address below.

Resources List

Useful addresses and sources additional to those listed in the Bibliography.

Association of Teachers of Mathematics (ATM),
Kings Chambers,
Queen Street,
Derby DE1 3DA.
Many publications, including: the journals *Mathematics Teaching* and *Micromath*; *Points of Departure 1* and *Points of Departure 2*; *Working Notes on Assessment*; *Working Notes on Microcomputers in Mathematical Education* etc. The ATM also published software and accompanying notes, e.g. 'SLIMWAM 1' and 'SLIMWAM 2' (Some Lessons in Mathematics with a Microcomputer), 'L – A Mathematical Adventure', etc. Also, local and national events for teachers.

GAIM Project (Graded Assessment in Mathematics),
King's College,
552 King's Road,
London SW10 0VA.

OCEA (Oxford Certificate of Educational Achievement),
University of Oxford,
Delegacy of Local Examinations,
Ewert Place,
Summertown,
Oxford OX2 7BZ.

The Mathematical Association,
259 London Road,
Leicester LE2 3BE.
The journal *Mathematics in school* is published five times a year. The MA also publish other journals, posters, and make available books from other sources (e.g. Marion Bird's book, cited in the Bibliography). Also local and national events for teachers.

Maths in Work Project (Mary Harris),
c/o ILECC,
John Ruskin Street,
London SE5 0PQ.
Many imaginative resources – *Wrap it Up* especially recommended for packaging activities and projects.

The School Mathematics Project,
Cambridge University Press,
Shaftsbury Road,
Cambridge CB2 2RU.
Many resources, including investigational material. Details and inspection material from this address.

Secondary Examinations Council,
Newcombe House,
45 Notting Hill Gate,
London W11 3JB.

Shell Centre for Mathematical Education,
University of Nottingham,
Nottingham NG7 2RD.
The packs produced in conjunction with the Joint Matriculation Board, and other publications.

The SMILE Centre,
Middle Row School,
Kensal Road,
London W10 5DB.

Order forms for all SMILE resources available from the centre. These are produced by teachers through the centre and include: classroom resources; microcomputer programs (MICROSMILE); the magazine *Investigator*; and the magazine *Splash*. (*Splash* is likely to be of interest only to SMILE teachers.)

Tarquin Publications,
Stradbroke,
Diss,
Norfolk, IP21 5JP

Publishers of *Dime* and *Leapfrogs* materials, a range of books, posters, puzzles, equipment and resources for practical and creative work.

West Sussex Institute of Higher Education,
Mathematics Centre,
St Michael's Building,
Upper Bognor Road,
Bognor Regis,
West Sussex

Many publications including: *Teachers Evaluating and Assessing Mathematics* (TEAM), produced in conjunction with the SREB.